MACHINE-AGE IDEOLOGY

THE UNIVERSITY OF NORTH CAROLINA PRESS • CHAPEL HILL & LONDON

Machine-Age Ideology

SOCIAL ENGINEERING AND

AMERICAN LIBERALISM,

1911–1939

JOHN M. JORDAN

The paper in this book

meets the guidelines

for permanence and

durability of the

Committee on Production

Guidelines for Book

Longevity of the Council

on Library Resources.

Design by Teresa Smith

Library of Congress Cataloging-in-Publication Data

Jordan, John M.

Machine-age ideology : social engineering and

American liberalism, 1911–1939 / by John M.

Jordan.

p.   cm.

Includes bibliographical references and index.

ISBN 0-8078-2123-3 (alk. paper)

1. Engineering—Social aspects—United States—

History.   2. United States—Social conditions—

History.   I. Title.

TA23.J67   1994

306.4'5'0973—dc20   93-2108

CIP

98 97 96 95 94   5 4 3 2 1

FOR MY PARENTS

CONTENTS

PART 4 RECONSIDERATION AND RETREAT, 1934–1939

ILLUSTRATIONS

ACKNOWLEDGMENTS

In the eight years I have lived with this book, it has led to debts—personal and professional, institutional and intellectual—of a magnitude I could never have imagined. To say that the following persons and agencies have enriched me is to underestimate completely the value of their contributions. In my defense, I can only repeat to them words of gratitude I recently read on the wall of a hospital: Words fail, but your consideration and professionalism did not.

While academics often feel underappreciated, librarians surely labor as hard with less reward. I have had the pleasure of working with the best in the business, including reference and interlibrary loan staffs at the University of Michigan, Michigan State University, and Harvard University (in ten different libraries). The same should be said for archivists at Columbia University, Cornell University, DePauw University, General Electric, Harvard Business School, the Herbert Hoover Presidential Library (especially Dale Mayer), the Rockefeller Archive Center (especially Thomas Rosenbaum), the Franklin D. Roosevelt Presidential Library, Smith College, Union College, the University of Chicago, the University of Illinois at Chicago, the University of North Carolina, and Yale University. Getting the photographs required special help, and my thanks are many to most of the aforementioned institutions as well as to Southern Illinois University, the National Bureau of Economic Research, and Stevens Institute of Technology.

I have received generous institutional support over the course of researching and writing. At the University of Michigan, the Horace H. Rackham School of Graduate Studies funded travel and photocopying. The Herbert Hoover Presidential Library Association, the Franklin and Eleanor Roosevelt Institute, and the Rockefeller Archive Center awarded me handsome travel grants to their respective institutions. A National Endowment for the Humanities grant allowed me to travel to the Howard Odum and Charles Beard papers. Most recently, the Milton Fund of Harvard University underwrote the cost of the photographs. To all of these agencies that invest in young and otherwise unproven scholars, my most sincere appreciation.

A number of scholars criticized this study's many and various drafts.

While I shamelessly took a lot of advice, most of these readers will still differ with me; the book's problems result from my hardheadedness rather than from their generosity. About a dozen anonymous referees for university presses and journals gave me encouragement, challenges, and advice, each of which helped in its own way. Steve Alter, Martha Banta, Paul Bernard, John Patrick Diggins, Samuel Haber, Brian Lloyd, Edward Shils, and Kevin Thornton read portions of the manuscript and offered their comments. Thomas Bender, Donald Fleming, Ellis Hawley, and Michael Scott Morton each read one entire manuscript or another; all of them reassured and motivated me. John Chamberlain, Martin Pernick, and Jim Turner sat on my dissertation committee, and years later I still find myself profiting from the wealth of their advice. Richard Marius read an almost penultimate manuscript and has turned out to be the best boss a young writer could ever hope to work with.

Special thanks must go to David Hollinger, the dissertation advisor who helped steer an inchoate conception to publication, and to Daniel Horowitz, who read the entire manuscript three times through. The former helped in ways large and small to shape the topic, expedite research, and sharpen my analysis, while the latter provided unfailingly constructive and perceptive criticism at important moments.

More generally, I must thank some colleagues who have discussed engineering, liberalism, and history with me and thereby shaped my thinking. The endnotes cannot do justice to the ways their approaches and ideas have permeated my own, but I can attempt to convey my appreciation to JoAnne Brown, Martin Burke, I. Bernard Cohen, and David Hollinger for enhancing this project. Similarly profound while less personal have been the influences exerted by James Kloppenberg, Dorothy Ross, Cecilia Tichi, and Robert Westbrook, all of whom forced me to solve the problem of engineering and reform in my own idiom.

For the gift—no other word will do—of sustaining intellectual friendship, I want to thank Stephen L. Keck, David Nichols, Bill Petersen, and Kevin Thornton. All of them bought me beers, pushed me to think, and believed in the project. I must single out Dave Nichols, who has read more of my writing than anyone else and who deserves medals for his endurance, humor, and syntactic tutelage.

Lewis Bateman, my editor at University of North Carolina Press, has demonstrated professionalism, savvy, and tolerance far beyond what any first author has a right to expect. Most important, he coaxed and prodded, never letting me neglect our shared standards and expectations for this project, which he has known almost since its inception. Working under his guidance has been a pleasure. His associates at the press, especially Ron Maner, were unfailingly cheerful and diligent as they schooled me in the mechanics of publication.

Like everything else I could attempt to give them, the dedication to my parents, Kurt and Shirley, merely symbolizes ongoing appreciation of their tireless efforts to raise me right. Their encouragement, judgment, and optimism continue to shape me, as does the example they have set as teachers. I can only hope to have my priorities as straight as they have theirs.

My in-laws, Douglas W. Bryant and Rene Kuhn Bryant, have impressed on me the persistence of tradition, the vitality of the written word, and the importance of not being too earnest. They also bore and raised a daughter to build upon their impressive legacy. Without Heather Bryant Jordan, there would be no book, and no reason or will to write one. She forced my thinking and writing back to earth when they got too ethereal, and renewed my spirit when it faltered. By her own scholarship, her smile, and her fierce and quiet inner strength she has enriched me and the book that I tried to write.

Concord, Massachusetts
March 1993

ABBREVIATIONS

The following abbreviations are used in the text. For additional abbreviations used in the notes, see pages 287–89.

ACCND	Advisory Committee to the Council of National Defense
AEA	American Economics Association
APSA	American Political Science Association
ASME	American Society of Mechanical Engineers
ASS	American Sociological Society
BMR	Bureau of Municipal Research
FAES	Federated American Engineering Societies
IGR	Institute for Governmental Research
IRI	International Industrial Relations Institute
LSRM	Laura Spelman Rockefeller Memorial
NBER	National Bureau of Economic Research
NESPA	National Economic and Social Planning Association
NRA	National Recovery Administration
NRC	National Research Council
NRPB	National Resources Planning Board
SSRC	Social Science Research Council
TVA	Tennessee Valley Authority
WIB	War Industries Board

MACHINE-AGE IDEOLOGY

INTRODUCTION

Your mission is to subjugate to the beneficent yoke
of reason the unknown beings who live on other
planets, and who are perhaps still in the primitive
state of freedom. If they will not understand that
we are bringing them mathematically faultless
happiness, our duty will be to force them to
be happy.

EVGENY ZAMYATIN, *WE*

For well over a century, Americans have idolized tech-
nology while chronically worrying about its implications. As one such wary
enthusiast concerned with this ongoing dialectic, I hope to contribute another layer of interpretation. Three main questions concern me: How do groups
define technology as appropriate or inappropriate? How does the portrayal of
technology influence our relation to it? and, ultimately, How does technology shape American life? In relation to these questions, this book explores
how technology, through a particular set of symbolic renderings, realigned
politics in the early and mid-twentieth century.

Previous contributions to this genre suggest Americans' simultaneous fascination and irritation with the power and disruption of applied science. In
the woods of Concord in 1844, Nathaniel Hawthorne contemplated his pastoral surroundings and was startled by "the whistle of the locomotive—the
long shriek, harsh, above all other harshness." The presence of what Leo
Marx has called "the machine in the garden" soon became a staple of American literature and cultural criticism. About a century later, James Agee and
Walker Evans memorialized the lives of southern sharecroppers in their
antidocumentary *Let Us Now Praise Famous Men*. While blasting Margaret
Bourke-White and other contemporaries with caustic criticisms, Agee and
Evans self-consciously failed to solve the problem of photographic intrusive-
ness; the moral implications of technological capability lingered. In 1969,
the triumph of the Apollo 11 moon landing stood in otherworldly contrast to

riots, assassinations, and widespread doubts about American involvement in the Vietnam war. Norman Mailer pondered a fire on the moon while both napalm and inner cities burned on planet earth. Fewer than twenty years afterward, what some regarded as a lame public-relations stunt—a program to launch a schoolteacher into space—tragically humanized the spectacular destruction of the *Challenger* space shuttle. NASA's "can-do" mentality, one icon of the space age, gave way to pathetic finger-pointing over the failure of a simple rubber O-ring.[1]

Each episode involves a telling juxtaposition of images—themselves technological artifacts—and each is fraught with political overtones. The federal government shapes much American technology by encouraging research, protecting intellectual property, taxing and regulating, and purchasing the culture's most advanced technology: weapons. The density of these many relations makes it difficult to understand (or reorient) technology, its symbolic representation, and its political components.

While few of us enter laboratories or even begin to grasp what research scientists do, automobiles, television, magnetic and optical storage media, telephones, and computers touch the lives of most Americans every single day. Air travel, xerography, and scores of medical technologies have also redefined contemporary life, yet how far have we analyzed these technologies' shock waves of impact? More insidiously, how do we comprehend and manage multiple understandings of technologies, rather than the technologies themselves? When the phrase "Do not fold, spindle, or mutilate" entered the lexicon during the 1960s, for example, many turned it into a symbolic protest of standardization and technological determinism. Because of the ubiquity of advanced technology in America, reflection on its place and ramifications has been slow; we laugh at *The Gods Must Be Crazy*, a movie about how a soft drink bottle changes "primitive" Africans, but essentially ignore what machines do to us—to our ideas and words and our politics. I argue that the symbolic and metaphorical understandings we make from our technics decisively shape our culture and institutions.

This book considers the multifold implications of engineering—rather, of some symbolic understandings of engineering—for reform politics and later mainstream liberalism. Growing numbers of middle-class managers, journalists, and academics looked to the tools of applied science in the search for a new political paradigm. The way they wrote and spoke about the state, the settings in which their discussions occurred, and the theoretical bases of their efforts all departed from previous practice. In the end, these reformers tried to gain political power by arguing that politics no longer existed; the methods and logic of applied science apparently guaranteed correct answers to every problem. Few grasped that the very process of determining the relevant questions was itself a political act. Because they had borrowed a seemingly perfect method, as the era's material environment bore witness, these individuals and their organizations could claim a "scientific" mandate to tell other people what to do, to force them, in the words of the Russian avant-garde novelist Evgeny Zamyatin, to be happy.

It is hard to blame these reformers for being dazzled, for they embraced a powerful force with apparently unlimited possibilities. Corporate capital, organized research, and applied science combined to demonstrate how humanity could understand and control the natural environment. During a structural transfiguration comparable to the Renaissance or any other cultural earthquake, citizens hailed one shining hero of the moment: the engineer.

With a sense of historical demarcation, people called this period the machine age. The introduction of household electricity, automobiles, refrigerators, telephones, and radio broadcasts completely altered the nature of the home, as power became available far from dams or other sources of generation. Mobility and communication could be undertaken at a moment's whim instead of by timetable, and these inventions quickly diminished the isolation of farms and suburbs from urban life. On a larger scale, ocean liners, commercial aircraft, and powerful locomotives advanced scheduled transportation, while such civil engineering triumphs as giant dams, bridges, and skyscrapers redefined public space and dwarfed human scale. Some of humanity's oldest dreams—powered flight and communications across distance—were realized. In its wake this storm of invention left a spirit of material progress, an inorganic machine aesthetic, and omnipresent talismans of applied science.

Dismissing the objections of the antimodernists and the abuses detailed by muckrakers, some Americans learned to believe in technology. (The hero of Eugene O'Neill's play *Dynamo* [1928] goes so far as to hurl himself into the generator-goddess to attempt divine consummation.) It takes an act of cultural amnesia to imagine how the promise of applied science could be so awe-inspiring that reformers sought to apply the lessons and principles of engineering to the ruling of America itself. Even though citizens have learned in the interim that innovation frequently imports new problems, that panaceas rarely pan out, and that the novelty of invention can quickly fade, the stunning degree to which these developments overhauled life and thought has yet to be completely appreciated.

Past accounts have called these reformers technocratic progressives, technocratic liberals, social engineers, and political rationalists. All of these designations fit to some degree, but they are also problematic. I use most of these labels at one time or another, but most frequently refer to this study's central figures as rational reformers, not to suggest that other critics were irrational but to emphasize their bedrock commitment to the power of reason. These Americans sought to remake their nation, to forestall radicalism on the left and plutocracy on the right, to encourage evolution instead of revolution. They wanted to escape political demagoguery and deadlock by invoking the method of applied science, convinced that it would lead to logical consensus from which purposeful action could proceed. At the same time, self-interest and an insufficiently critical attitude toward authority made their attempt to circumvent and reinvent politics inevitably and inherently political.[2]

These people who considered America in the terms of controlled cause and effect—the basis of engineering—did so in the face of several kinds of opponents. After 1880, increasing ethnic diversity changed the face of politics, in cities especially. Some reformers' reliance on technical rationality and the experts who applied it was undoubtedly a response to the perceived dangers of mass democracy, as the fallacies of the eugenics movement would suggest. In addition, women recently granted suffrage threatened the power of a traditional elite. Rational reforms also served to screen out women, who were not well represented in the academic, engineering, or philanthropic professions where social engineering took hold fastest and most firmly.

Rational reformers used several key words to describe their opponents. To be *radical* was to challenge the sanctity of private property and to threaten the existing custodians of material wealth. No matter how scientific the reformers claimed to be, few objectively assessed the implications of capitalism and instead took it as a given, not a political outcome. The other key word was *emotional*. To Anglo-Protestant men in an expanding middle class, black southerners migrating northward, Irish-American political empires, and women generally thought to be too sentimental for public responsibility constituted threats from several sides. Still, it appeared that the center could hold in defiance of "untrained and unchastened uplifters," as one technocrat called his imagined opponents, or in spite of those groups Charles Beard named as roadblocks on the highway of progress: "economists, politicians, statesmen, labor leaders, and feminists." Howard Odum contrasted the "scientific-liberal" view of those in white hats to other outlooks: the "dogmatic-conservative," the "emotional-radical," and the possibly Nietzschean "agnostic-objective."[3]

In this outlook, the antithesis of emotion was of course reason, the stuff of science and the source of the modern world's mechanical marvels. The reformers understood rationality in simplistic terms, however, often arguing for the existence of exact answers to all questions. Science's methods and its spirit of inquiry could solve social problems just as engineers could calculate correct load factors for bridges or lift coefficients for airplanes. Debate, compromise, and negotiation would thereby be streamlined. But few of these reformers realized that they merely had substituted one system of belief for another; in so doing, they attempted to win political contests by denying the legitimacy of understandings other than their own. Grammars and logics originating in alternate readings of experience and based on the family, the body, the jungle, or the church were ignored or dismissed as antiquated. For some Americans, the scientific worldview triumphed without question.

When politics is defined broadly as the pursuit of authority within a group, and not merely in terms of formal governance, the place of language as cultural currency becomes especially relevant. Metaphors, in particular, function as means of often artificial agreement; each hearer of a given metaphor carries a private understanding that may be at odds with others. What JoAnne Brown has called "the seemingly objective logic of its literal referent" allows metaphor to create a frequently illusory consensus that would be impossible if participants had to reach explicit agreement on definitions. Like Archimedes, people search in vain for a platform apart from discourse from which to apply leverage to their world. Language, especially metaphor, thus operates as an ongoing epistemological contract that, because it is implicit, is rarely contested. The meta-language of politics determines many of its outcomes; those who define the terms usually win the debate. The interwar era illustrates particularly well how the epistemology of politics interacts with the politics of epistemology.[4]

My focus on the rhetorical aspects of rational reform will center on the phrases connecting social institutions, cultural process, and political practice with technological innovation and performance. Words and phrases such as *efficiency*, *machine age*, and *planning* carry particular importance because their imprecise and wide use allowed them to be interpreted in a variety of ways. A most significant phrase was *social control*, a vaguely defined objective: was society to be controlled by some elite, or could society control its collective destiny in contrast to drifting with the tides? Reformers frequently used the term as though it meant the latter, even while working toward the management of the many by the few.[5]

In contrast to these often unreflective political languages, rational reformers devised some coherent theories to justify their efforts. While Europeans from Francis Bacon to Auguste Comte to John Stuart Mill had considered the governance of the state in terms of mechanical and scientific apparatus, these foreign conceptual influences seem negligible in the American context. Instead, homegrown political economists such as William Graham Sumner and Lester Frank Ward analyzed society with words derived from the realm of scientific inquiry, while the popular author Edward Bellamy designed a technological utopia premised on machine productivity and industrial logic. Many intellectuals in an increasingly secular and scientific age found Judeo-Christian metaphysics less and less satisfactory as a source of confidence and stability. In response, philosophers and social theorists in the early twentieth century gave closer consideration to the sensory and intellectual appropriation of—and control over—the natural world. Most notably, John Dewey and Thorstein Veblen made empiricism, as they understood it to function in natural science, the foundation of their distinct but related theories of cultural organization and political change.

For these theorists and their many followers, America could be seen as an increasingly incorporated entity; productive and communications technology helped to make both industrial enterprises and large political institutions more robust after the Civil War. Some twentieth-century intellectuals began to assert that such a national, interconnected America could best be understood in terms of proximate physical causes. Older ideologies—based in religious authority, traditional mythologies, or pioneer adaptation—fell from favor. Veblen and Dewey differed over the implications of this newer mode of political analysis. Both, however, thought that industrial America needed rulers adequate to its new complexity. Other writers concocted similar theories. In 1909, Herbert Croly addressed the increase in productive scale in *The Promise of American Life*. He argued that skilled administrators had to understand and rule the nation as a whole entity because of its economic and technological coalescence. Advocates of the technocratic strand of Progressive reform implemented these ideas in systematic studies of political administration, managerial innovation, and institutional economics between 1890 and 1910. Such empirical investigations allowed intellectuals to react to challenges of scale, power, and complexity posed in a multiethnic, geographically dispersed, and technologically sophisticated United States.[6]

The theory behind social engineering, artificially simple in its logic, drew upon intellectual concomitants of industrial might. Social problems in a technological age, the reasoning went, were of a different order and magnitude compared with what had confronted the reformers' predecessors. Politics as a governing device had become outdated, falling prey to the mass appeals and backroom deals frequently thought to characterize it. Labor unrest, crime, and poverty were thus seen not as moral problems but as managerial ones. Empirical studies to document the magnitude and locus of a given ill could be followed by equally empirical efforts to solve it. The same methodology that enabled steel to be manufactured to previously unattainable degrees of hardness could solve ethnic tensions or alleviate poverty. Scientific management and public administration—new fields that appeared during this period—thus shared an outlook: practitioners' jobs consisted not of fomenting consensus or defining goals, but of troubleshooting and problem solving. Such an approach held intellectual appeal because pragmatic standards for justifying action centered on performance alone. Croly, Dewey, and the managerial innovator Frederick W. Taylor thus reinforced each other and encouraged others to pursue similar lines of argument.

In addition to linguistic and theoretical issues, this book addresses institutional change. One of the earliest managerial organizations, the Taylor Society, provided an initial home for many who applied engineering understandings to society. During the 1910s, intellectuals associated with the *New Republic*, foremost among them the young Walter Lippmann, also endorsed similar precepts. World War I mobilized many engineers, academics, and managers, giving them a brief but often tantalizing experience with large-scale rational social governance. Afterward, Herbert Hoover, the "Great Engineer," continued to lead efforts in this direction. He was ably assisted by a new breed of foundation administrator, who like the *New Republic* intellectuals worked primarily in New York. The Rockefeller philanthropies in particular funded many efforts to apply scientific rationality to social problem solving. Cross-fertilization within research staffs, boards of directors, and government offices enabled a cadre of men and women who shared a language of reform to practice at the highest levels of politics what had been only theory a generation before. Many of these enterprises—the Social Science Research Council, for example—extensively debated the application of scientific modes of reason to a democracy, and this self-consciousness helps to illustrate for later generations the tangle of motivations at work in this sector. In the early years of his presidency, Franklin Roosevelt relied heavily on social scientists from these institutions. Even so, their limited success and the basic ideological and programmatic schizophrenia of the

New Deal made "planning" little more coherent, or successful, than "efficiency" had been twenty years earlier.

The history of the rational reformers, and the terms they employed, can tell us a great deal about liberalism, the code with which most of these intellectuals aligned themselves after 1920 and still one of the most misunderstood words in our lexicon. In the indistinct but crucial realm of political culture, the engineering and managerial influence persisted well after World War II, finding its highest expression in the 1950s and 1960s, when corporate managers controlled important sectors of the federal government. And in the Kennedy and Johnson administrations especially, social scientists shaped policy in the educational, social, economic, legal, and international realms, often using mathematical models. Following the lead of their prewar forebears, postwar liberal theorists denigrated the sacred and hailed the secular.

A discussion of the "best and the brightest" of the Great Society must begin with what Taylorites called "the one best way" in the first years of the century. In his study of post–World War II America, Godfrey Hodgson argues that the midcentury "liberal consensus," while not a clearly defined ideology, did stand on six interrelated presuppositions, several of which derive specifically from the outlook of the social engineers discussed here. Supreme confidence in American productive capacity and the concomitant permanence of economic growth (Hodgson's maxim number 2) descended from Taylorism and managerialism; the belief that "social problems can be solved like industrial problems" (number 4) explains itself. The implications of rational reform linger into the present. As Alan Brinkley recently pointed out, the popularity of Ross Perot's 1992 presidential bid relied on both his association with technocratic solutions and his belief in a resurgence of American productive capacity.[7]

The reformers' fascination with scientific method, machine process, and large-scale managerial organizations as analogues for government cannot, however, be viewed only in terms of its painful consequences. Without a doubt, engineering's inevitably hierarchical logic threatened the delicate balance of democratic politics, and the hubris implicit in any attempt to win a battle merely by declaring an opponent's ideas outmoded and trivial also irritates. Still, few of these figures can be easily dismissed; merely reading technocratic authoritarianism backward to some relevant predecessors oversimplifies both politics and historical causation. It is imperative to understand the men and women in this book in the hopeful terms that motivated them. Most had some genuinely humanitarian aspirations, even if they did misread both engineering and politics. The state was nebulously compared with bridges, dynamos, and ships, while the term *engineer* could generically connote inventors and scientists as well as professional applied scientists.

In the end, the actual process of applied scientific innovation changed the world more quickly than the reformers could. (It may be that industrial utopians like Bellamy and Charles Steinmetz, who intermingled political theory with technological forecasting, understood the future better than they realized.) By World War II, the machinery of mass production and monuments to civil engineering, while still important, became less useful as cultural symbols. Instead, television and other technologies of communication accentuated what the rational reformers opposed: emotional appeals to irrational mass desire. In addition, the medium of communication itself changes the nature of the message being carried; imagine William Jennings Bryan with a microphone, or Abraham Lincoln with a teleprompter. Political referents and rhetoric that may have cohered in a town meeting or stump speech often failed to persuade when converted into magazine advertising, moviehouse newsreels, or radio addresses.[8]

In response to technological innovation, the rational reformers all tried to invent industrial-strength tools of "social control." They often crossed existing political lines—between capitalism and socialism, labor and management, Democrat and Republican, public sector and private—trying to replace contentious and apparently impotent political devices with more "scientific" arrangements. Their solution to the problems of politics was thus an antipolitics, an attempt to find a method whereby efficacious action could proceed. Goals for these methods often went undefined, as did the politics of choosing among different methods. Accordingly, while the so-called social engineers operated on the basis of complicated motives—some noble, many not—an uncritical appropriation of engineering as myth, method, and metaphor for reform was their crucial mistake. It does matter, however, that they asked many of the right questions.

Despite the Byzantine story that one could make of rational reform, the political theorist Hans Morganthau succinctly addressed the crux of the matter. "An age," he wrote, "whose powers and vistas have been multiplied by science is liable . . . to exalt in the engineer a new man whose powers equal his aspirations and who masters human destiny as he masters a machine." While many Americans clearly indulged in such exaltation, they did so for complicated reasons and with mixed results. I come neither to praise nor to condemn these people, but to comprehend them. They saw correctly that the industrial age tested government with new and intractable technical problems that required efficient management. But the modern age also presented social stresses requiring artistic political attention, and the fixation on administrative technique distracted energy from other no less necessary governmental functions. In the end, inventors and engineers redefined politics without planning to, while social engineers never built the rational republic they foresaw. This same paradox born of political striving and technological capacity continues to confront us: we still seek to reconcile efficiency with justice, performance with compassion, and competence with statesmanship.[9]

PART ONE ···

PREDECESSORS
··

1 8 8 0 – 1 9 1 0

You see, getting down to the bottom of things, this

is a pretty raw, crude civilization of ours—pretty

wasteful, pretty cruel, which often comes to the same

thing, doesn't it? . . . Our production, our factory

laws, our charities, our relations between capital

and labor, our distribution—all wrong, out of gear.

We've stumbled along for a while, trying to run a

new civilization in old ways, but we've got to start to

make this world over.

THOMAS EDISON

1

ORIGINS OF AMERICAN RATIONAL REFORM

The politics of efficiency, social control, and planning originated in tangible causes and effects, not in mass movements or charismatic leadership. Even the philosophical bases of social engineering—pragmatism and Veblenism—begin with human action and emphasize performance. The rational reform impulse stressed present-tense problem solving, not historical precedent. European predecessors like Comte and Mill influenced a few important individuals—especially Herbert Croly, who was actually baptized into Comte's religion of humanity—but most Americans tried to redesign society with little sense of intellectual genealogy. Antitheoretical theory begat apolitical politics.[1]

Rational reform drew its vigor from intellectual, professional, and material sources: Lester Frank Ward and Thorstein Veblen, academic social science, and engineering successes. In each instance, social change hinges on the appropriation of apparently scientific technique rather than on virtue, votes, or received wisdom. These innovators influenced later generations to continue to flee social ideology and personal metaphysics toward scientific control and existential certainty. Within both theory and practice in late nineteenth-century America, similar themes reappear, always grounded in the ever more evident power of applied science.

WARD, VEBLEN, AND THE SCIENCE OF SOCIETY

Lester Frank Ward, who anticipated many aspects of rational reform, differed markedly from later generations in his firsthand knowledge of natural science—he published several volumes of botanical and geological material and worked with John Wesley Powell as a paleobotanist. His acquaintance with scientific progress made him revise Comtean positivism to keep pace with the promising developments of the late nineteenth century. By understanding "the operations of a state" as "natural phenomena," Ward could begin the move toward a theory of social engineering. After politics was viewed as nature, it could be manipulated to fit human design: the "inventive stage embraces the devising of methods for controlling the [social] phenomena so as to cause them to follow advantageous channels, just as wind, water, and electricity are controlled." Ward's linkage of scientific inquiry to control influenced a significant body of twentieth-century social thought, but few of his contemporaries.[2]

Ward substituted a scientific (in the Comtean sense, a positive) understanding of human agency for William Graham Sumner's Darwinian combative randomness. In Ward's theoretical state, science would enable citizens to differentiate themselves from animals by the application of knowledge. The beginnings of political engineering appear in Ward's earnest prose of 1893: "Every wheel in the entire social machinery should be carefully scrutinized with the practiced eye of the skilled artisan, with a view to discovering the true nature of the friction and of removing all that is not required by a perfect system. . . . The legislator is essentially an inventor and a scientific discoverer." Note that the word *engineer* never appears; Ward called on the skilled artisan as his ideal. Before social engineering could become a possibility, reformers needed living examples of empirically based control over natural forces.[3]

. . . .

"It is what they used to burn folks for." So wrote Ward of Thorstein Veblen (1857–1929), the American thinker who most completely challenged the status quo by exploring the future of the technical state. The essential elements of his concept of society—an anthro-utopian world of consumer plenty, rational technique, and demystified authority—appear repeatedly in the works of later followers. Veblen's political thought connected the nineteenth-century utopian tradition to the empirical social sciences of the early 1900s.[4]

It is initially useful to consider Veblen in relation to the pragmatic tradition. He studied under Charles Sanders Peirce, worked alongside John Dewey and George Herbert Mead, and read William James closely. Sharing the pragmatists' stress on truth found in meaningful human action, Veblen named the tendency toward usefulness the "instinct" of workmanship. He defined human life as a series of causal actions, the meaning of which is found in their effects. Veblen also drew upon Darwinian science, aspiring to analyze society as an evolving set of institutions and processes—at times the words appear to be interchangeable—where survival is proof of exhibited fitness; he wrote that "the evolution of social structures has been a process of natural selection of institutions." In Veblen's view, science encouraged "matter-of-fact" habits of thought: the scientist sought to analyze a situation in terms of strictly observable cause and effect, not progress toward a far-off goal.[5]

Despite assuming the pose of the scientist, Veblen created in his economic anthropology not so much a science as an epic allegory. The Norse sagas he so admired exemplify the scope, moralism, and poetic license that Veblen mimicked in his own writing. His timetable of human events was, at best, hypothetical, even given the state of academic anthropology at the turn of the century. When he wrote of "an unbroken cultural line of descent that runs back to the beginning," Veblen operated not on the evidence of field studies, archeological digs, or linguistic analysis. The construct began with the "golden age" of savagery, in which humanity was peaceable and cooperative and which functioned in much the same way that the fictive state of nature did for Locke, Hobbes, and the other contract theorists. Because the governing factor in Veblen's theory of cultural development is the state of man's technology, he posited that with the advent of new tools of killing, acquisition by seizure implied the origins of private property, and eventually the state evolved to protect property rights. As industrial technology improved, however, human institutions were always in arrears; never was a given cultural arrangement adequate to the capabilities of current tools and techniques. Thus, for Veblen, adjustment is a primary value: always his critique of culture is aimed at "archaic" institutions inadequate to current exigencies. Change—in a Darwinian sense, never in the process of reaching teleological goals—was the solitary imperative.[6]

Because the species possesses an instinct of workmanship, people can change their world through the discovery of new technology. "Man's great advantage over other species in the struggle for survival," Veblen wrote, sounding a lot like Ward (whom he cited), "has been his superior facility in turning the forces of the environment to account." Veblen replaced "economic man" with another fiction—man the worker—who retained the instinct of workmanship which "disposes men to look with favor upon productive efficiency and on whatever is of human use." In his essay of 1898 on the topic, Veblen elaborated: "*All men* have this quasi-aesthetic sense of economic or industrial merit, and to this sense of economic merit futility and inefficiency are distasteful. In its positive expression it is an impulse or instinct of workmanship; negatively it expresses itself in a deprecation of waste." Before phrases like *home economics* and *social efficiency* captured cultural aspiration toward techniques of political renewal, he constructed an illusory anthropology embodying the efficiency criterion.[7]

Veblen argued that because culture begins with the advancement of its tools, "the scope and method of modern industry are given by the machine." Indeed, he made industrial development the raison d'être of human societies; for a thinker who opposed teleology so strongly to espouse technological fetishism is but one of the puzzles of Veblen's work. "The collective interests," he wrote, "of any modern community center in industrial efficiency." But instead of meeting material needs with efficient production, capitalist industry was marked by personal, qualitative, and status-conscious habits of thought. The "pecuniary" mindset adopted by the captains of finance and industry overruled the impersonal, quantitative, and use-conscious mind "disciplined" by the machine. In other words, the expression of the instinct of emulation, in large measure through competitive display, negated the impact of the instinct of workmanship encouraged by the machine process. Even though science enabled humanity to shed archaic rituals and beliefs, the pecuniary instinct denied the industrial imperative to make goods, leaving Western culture to lag further behind the rapidly advancing state of the industrial arts.[8]

The growing cultural authority of the scientific method appealed to Veblen, whose substitution of an allegedly scientific rationalism for a religious one foreshadowed similar developments within social science. In both modern technology and modern science, he wrote, "the terms of standardization, validity, and finality are always terms of impersonal sequence, not terms of human nature or preternatural agencies." A movement toward "precise objective measurement and computation" discounted "postulates and values which do not lend themselves to that manner of logic and procedure." The empirical scientist stood as the final authority in such a culture. Accordingly, Veblen adamantly encouraged the abandonment of the conveniently vague metaphors of classical economics—which allowed the construction of theories "without descending to a consideration of the living items concerned"—in favor of empirical methods; glorification of some literally invisible hand should, he contended, give way to examinations of concrete relations of exchange. His own work, however, relied only rarely on precise statistical data, leaving students like Wesley Mitchell and Robert Hoxie to the mind-numbing plug-and-chug of rigorous quantitative analysis. Veblen, meanwhile, continued his unsystematic but suggestive reasoning.[9]

Not only did "opaque cause and effect" generate an ethics for Veblen, it was his metaphysics as well. Despite disclaimers about a "morally colorless" standpoint defined by scientific observation and logic, the very survival of materialist reasoning proved its evolutionary fitness. Anything "not consonant with these opaque creations of science is an intrusive feature in the modern scheme, borrowed or standing over from the barbarian past." As frequently happens in his writing, the letter of the text must be distrusted: "The machine process gives no insight into questions of good and evil, merit and demerit, except in point of material causation." Here the "except" is precisely the point: the very logic of the machine, built on a chain of causal sequences, contains its own moral imperative. With knowledge linked to control and inquiry tied to application, Veblen located moral perfection in mastery of causal sequences, in process rather than in teleology.[10]

How would such reasoning affect politics? Veblen wrote relatively little on the topic, for his was not a particularly programmatic social criticism. Citizens supported the state, he contended, for two reasons: patriotism and profit. In keeping with his habit of damning the archaic, Veblen argued for an industrial government, one able to curtail pecuniary tendencies, "absentee ownership" in particular. Self-proclaimed political scientists would seem to be logical inheritors of his mission, but in 1906 he called the discipline only a "taxonomy of credenda," a particular insult because taxonomies were a legacy from pre-Darwinian science; their static analyses failed the test of evolutionary capability. Because Marx was handicapped by a have/have-not dichotomy, he too lost favor. Veblen's insistence on a breakdown between pecuniary (money-making) and industrial (goods-making) pursuits led him to consider industrial socialism. In 1893 he suggested that the "whole trend of the modern industrial development is distinctly socialistic." By 1905, Veblen would write of socialism, possibly a non-Marxian variety, as the "manifestation of machine-thinking for politics."[11]

The evidence for Veblen's conventional socialism remains inconclusive, however, insofar as he never clarified his terms or committed himself to existing movements. Instead, his politics relate closely to the utopian tradition. He studied Henry George while an undergraduate, translated Ferdinand Lassalle's *Science and the Workingmen* for the socialist International Library Publishing Company, and read Edward Bellamy's *Looking Backward* aloud with his wife. While he later attacked single-taxers and other believers in "cultural thimblerigs" (shell games), Veblen never used Bellamy's concept of nationalism in any but a favorable context. Veblen's world of maximized production, noncompetitive consumption, and demystified bases for belief and action closely resembles the world of 2000 in *Looking Backward*. Consider Veblen's view that socialism is "but the logical outcome" of evolving democracy in the modern age; Bellamy had argued that a bloodless, logical transfer of power would begin the utopian age. Veblen's approving use of Bellamy's term for socialism—"the Nationalist state"—also had to be deliberate in an age of Bellamy clubs and other efforts to make real the promise of the book. Or compare Veblen's paraphrasing of socialism as "the industrial organization of society" to the primacy for Bellamy of the "industrial army."[12]

In such a world, rationalized allocation of goods would lessen competitive displays of property and free much of the work force for production of more essential goods. This similarity involves the core of both Veblen's theory and the appeal of Bellamy's utopia. In addition, Veblen's dismissal of national boundaries in both *The Theory of Business Enterprise* and his World War I writings mirrors Bellamy's espousal of the popular belief in a "loose form of federal union of worldwide extent." Finally, the categories of waste outlined in *The Engineers and the Price System*—salesmanship, production of superfluous goods, systematic dislocation due to conventionally misguided business strategy, and unemployment of men, materials, and equipment—follow Bellamy's categories. He had pointed to "mistaken undertakings," "competition and mutual hostility" within industry, "periodical gluts and crises," and "idle capital and labor" as prime causes of waste. Now dismissed as oppressive yet sentimental, Bellamy influenced the social thought of the next half-century to a degree as hard to imagine as it was significant.[13]

Veblen literally looked backward to the Viking saga for his lost ideal society. Unlike the arts and crafts movement, which he attacked for finding "salvation" outside the machine process, Veblen wanted to retain that process because it discouraged the institution of private property. When industrial society could produce so many goods that scarcity would no longer create emulative value, and when humans no longer felt the need to compete through possessions (or in any other medium), meaningful community would return. "The most ancient and most consistent habits of the race," he wrote, apparently discarding his condemnation of archaic traits, could return, and "the ancient bent may even bear down the immediate conventional canons of conduct." Later in his life Veblen appears to have backed off some of his Bellamyite conceptualizations, but he retained his belief in the social good of maximum industrial production.[14]

Veblen had a wide and profound impact on American thought, in social engineering circles especially, in the first half of the century. The works of Wesley Mitchell, John Kenneth Galbraith, Lewis Mumford, Walter Lippmann, and Max Lerner bear his mark. Adolf Berle and Gardiner Means began their monumental study *The Modern Corporation and Private Property* with a footnote to Veblen. Malcolm Cowley and other *New Republic* editors found in their informal poll of "books that changed our minds" in 1939 that Veblen's name came up more than any other (Charles Beard was a safe second, followed by Dewey, Freud, Spengler and Whitehead in a tie, and Lenin in seventh). *The Theory of the Leisure Class* was the most frequently mentioned mind-changing book in lists from such intellectuals as Kenneth Burke, Robert Lynd, and Thurman Arnold. Rexford Tugwell wrote the Veblen essay in a collection devoted to poll winners.[15]

What sorts of judgments might the late twentieth-century reader render? Balance is difficult, for Veblen's strengths are solidly embedded in weaknesses. His prose style delights when it does not infuriate; there is real malice in such pointed humor. Allegorical anthropology makes for grand epic but poor scholarship. His most penetrating insights are the most densely obscured. The utopian future lies in a return to the Viking past. Intellectual sloppiness frustrates all the more coming from a mind of such brilliance. Breadth of learning teeters on the precipice of dilettantism. Like another long-winded moralist, Reinhold Niebuhr, Veblen could turn phrases of arcing power and elegance while frequently being unable to edit his way out of his beloved ironies. Perhaps what is left is a tribute to his supreme intellectual arrogance: we are forced to confront the prophet on his own mystical territory, regardless of our discomfort in doing so.

Whatever its weaknesses, Veblen's work foretold the future. By 1911, the efficiency craze began to bring to public attention the drive toward technique as a nostrum for all manner of lacunae in American values and, ever less implicitly, politics. Veblen's antipathy toward teleology prefigured a declining interest in broad social goals as progressively improved means were thought to be ends in themselves. As causation implied its own morality for Veblen, so did improved techniques imply worthy purposes for the rational reformers.

In the late nineteenth century, engineering was a young, growing, and mis-understood profession. The white-jacketed research scientist so familiar to twentieth-century observers had not yet captured the public imagination, so that science and engineering were distilled into the figure of the inventor and, later, the engineer. Balkanized by specialty in a way the American Medical Association, for example, never was, engineering in the 1890s began to define its distinct professional identity: societies published jour-nals, colleges standardized curricula, and meetings and annual conventions became more common. By the early twentieth century, Ward's industrial ar-tisan quickly became a curiosity. Meanwhile, the professional engineer, who united the practical ingenuity of the Yankee mechanic with the theoretical rigor of modern science, achieved the status of cultural savior, figuring prominently in advertising, fiction, and the career aspirations of young men.[16]

While becoming increasingly important to an industrial economy, the pro-fession had to cope with rapid expansion and visible success. In 1880, all American engineers numbered only 7,000, while forty years later 136,000 practitioners defined themselves with professional specializations. (By 1935, real engineers had differentiated themselves with 2,518 different job titles. Pretenders—including a matrimonial engineer, a hot-dog engineer, and a touchdown engineer—also used the term.) As of about 1910, engi-neering groups attained professional stature by tightening their grip on spe-cialized knowledge, implementing standardized procedures, and communi-cating in a newly private vocabulary. Paralleling these internal tendencies, American readers developed an appetite for engineering as the stuff of art and hero worship. Engineers appeared to insure profitability without being mercenary and to pursue truth without being impractical, incorporating the twin heritages of business and science. Noting the capacity of engineers to dream and to create, observers celebrated them as "true poets, makers whose creations touch the imagination and move the world." Of course, not all of this acclaim applied to professional engineers alone; the name Edison, for example, instantly connoted inventiveness. Even though the disruption of supercharged industrial expansion led few public accounts to respect pro-fessional engineering's self-definition, the vaguely drawn engineer became a powerful cultural symbol.[17]

Plentiful artifacts embodied this celebration of control, inventiveness, and efficiency. It must be stressed, however, that engineering was never accorded the status of a classic profession. While they compared themselves to physicians and attorneys, most engineers remained employees of large businesses; they never won sufficient autonomy to control the market for their services. Despite this distinction, the engineer shared in the widespread glorification of industrial America. The urban landscape, Erector Sets, and reams of advertisements proclaimed engineering's many socially redemptive characteristics. Engineers appeared as heroes in over one hundred silent movies and in novels that sold roughly 5 million copies between 1897 and 1920. The men Herbert Hoover called "new intellectual engineers" surveyed many possibilities.[18]

Consider how the profession presented itself in a book entitled *Engineering as a Career* (1916). A distrust of rhetoric, political or otherwise, comes through immediately, for the engineer's "language is one of blue prints. He does not talk in words." Instead, technical experts claimed moral integrity and even virtue: "The greatest strength of an engineer is the justice of his position." Such authority resulted from an "honesty" deriving from the realization that "they cannot juggle with Nature." This reverence for facts and forces contributed to a view of the world that echoed that of the nineteenth-century cleric, but with an updated relevance: the engineer, the authors claimed, "is religious" even though "he may have no creed." Implicit trustworthiness, unlabeled virtue, and proven efficacy became common themes in discussions of the profession.[19]

Certain engineers added hopes for the scientific transformation of society and politics to more conventional aspirations toward professional recognition. The mechanical engineer George Melville claimed much for his calling. When a future historian describes the United States at the beginning of the twentieth century, he wrote, "and attempts to award the credit for the existing comforts and conveniences, the major part must be given to the profession of engineering." Another colleague agreed and made similar claims: "Engineering creates modern life. Take away engineering, and we are what our predecessors were. Add engineering, and the modern world is." Yet the rewards for such a dramatic transformation came too slowly for some tastes. Melville asked, "What does the engineer owe society when society owes so much to the engineer?" He continued, wondering why engineers had not received as much credit as they thought fitting: "Is the reason not very largely because the engineer has hitherto been content to do the work and then fade into the background, leaving the talking and the management to the lawyer and the politician?"[20]

One solution to what Talcott Williams called "the disadvantage of a silent profession," then, was for engineers to involve themselves in civic affairs. Williams told a meeting of mechanical engineers in 1908 that "in every social system . . . a calling will have weight, influence and authority in proportion to its access to the center and origin of authority and power." Engineers had to gain public esteem for their recommendations to gain credence, for "failing to get credit, he fails to have weight with the community, which blunders and wanders for lack of his guidance." Melville, too, called for this merger of public service with self-service: "The engineer himself should take a vital and directing part in the administration of affairs." Despite his other "engrossing and exacting" duties, "where it is a matter of self-interest, the engineer, like other men, can find time for this extra work." Melville argued that not until the engineer promotes "the highest efficiency of the Government can he truly say that he is, in the fullest sense of the term, a good citizen of the Republic." If citizenship as virtue no longer motivated civic behavior, if indeed it ever did, public-mindedness still could have its tangible benefits.[21]

Some engineers were in fact making inroads into politics. In Schenectady, New York, Charles Steinmetz of General Electric joined the Socialist administration of Mayor George Lunn as a member, and later president, of the board of education in 1911. (Lunn's administrative assistant, for a time, was Walter Lippmann.) Elsewhere, managerial reformers later known as efficiency experts or efficiency engineers pursued administrative goals compatible with the material achievements of their industrial counterparts. One of these experts was Morris Llewellyn Cooke, a mechanical engineer who attempted to distribute engineering's benefits to a wider public. A close associate of the managerial innovator Frederick W. Taylor, Cooke merged managerialism and reform from the early 1900s well into Franklin Roosevelt's administrations. He initially worked within the American Society of Mechanical Engineers to increase its civic awareness, giving a paper to the 1908 convention encouraging engineers to consider social issues, and not merely the concerns of the firm. Cooke put his beliefs into practice when Rudolph Blankenburg, the reform-minded mayor of Philadelphia, invited him to reorganize the city's public utilities in 1910.[22]

Efforts to involve engineers in politics cohabited easily with another early twentieth-century enthusiasm: the infatuation with Theodore Roosevelt's resource managers who pursued the conservation ideal. Analyzing the efficiency crusade a few years later, Taylor invoked Roosevelt's efforts as the standard by which successful movements were measured: "The interest now taken in scientific management is almost comparable to that which was aroused in the conservation of our natural resources by Roosevelt." Taylor also began *The Principles of Scientific Management* with a "prophetic" quotation from Roosevelt: "The conservation of our national resources is only preliminary to the larger question of national efficiency." The two programs shared many features. Indeed, the middle-class reformers involved in resource movements inclined much more toward managerialist capitalism than Thoreauvian nature worship. By 1908, the conservation ideal connoted a broad-ranging program of rationalization among a small but influential constituency.[23]

Several factors made the Roosevelt conservationists precursors, not prototypes, of technocratic progressives of the World War I period. First, the competence of the technicians called upon to run the state increased dramatically between 1910 and 1920; social science, political administration, and managerial capitalism all grew stronger in the 1910s. Roosevelt also maintained an ambiguity about means and ends that dissipated within technocratic progressivism, giving way to an unabashed concentration on the means of "scientific" social reform to the exclusion of ends. He sought the advice of experts, believing it possible to rationalize the state and transcend power politics; he maintained a robust faith in applied science. The White House Conference idea, the commission concept, and the impact of his own Harvard education testify to this concern for enlightened expertise. While Roosevelt looked forward to applying the same approach to nonresource issues like labor relations, conceptions of good and evil remained the final standards of judgment; he wrote that he was "disposed to interpret economic and political problems in terms of moral principles." For inspiration Roosevelt looked not to the technician's futuristic industrial state, but back to a less class-bound agrarian society where he thought personal virtue had been given freer expression. Thus Roosevelt pursued both nineteenth-century moralism and twentieth-century managerialism, held in tension by the enormous appeal of his personality and his considerable ego.[24]

SOCIAL SCIENCE PROFESSIONALIZATION

The profession that would be home to many social engineers of the 1920s entered a formative stage before 1910: the academic social sciences defined themselves in this period. But most members of the American Economics Association, the American Political Science Association, and the American Sociological Society faced competing pressures. The ways that the professors and their superiors resolved these tensions shaped the profession's intellectual and political agendas in the Progressive period and after. Social scientists forced to reconcile involvement with detachment and scholarship with impact looked to engineering, clerical, and other predecessors, but in the end they had to chart their own course.

The role of moral exemplar, partially filling the void left by the decline in the prestige of the clergy, appealed to social scientists for two reasons. Many sociologists, especially, of the very early twentieth century retained deep roots in nineteenth-century moral reform movements. In addition, these new professionals aspired to a style of scientific practice that became invested with images of purity and nobility in the late nineteenth century and thus encouraged secular piety. Their definition of professionalism thus substituted the morality of presumed objectivity for the morality of ethical opinion.[25]

In contrast, university presidents wanted to show off useful results in fields concerned with day-to-day modern life. In his presidential address to the third annual meeting of the APSA in 1906, Albert Shaw addressed the tension. He paid his first respects to objectivity: "This Association is not partisan, is not a body of reformers, is concerned with no propaganda." But neither was the APSA detached from issues: "Above all, it can help bring to [public affairs] . . . the spirit of calmness, of inquiry, of reasonable discussion—in short, the scientific spirit." As Progressive reform became an acceptable middle-class political response to a changing economy and society, advocacy of a method or "spirit" (as opposed to advocacy of a goal) by social scientists was allowed and, in the famous Wisconsin Plan, encouraged.[26]

The appeals of objectivity and advocacy competed as professionals sought to empower themselves. Organized into groups paralleling labor unions, horizontally integrated corporations, or merchants' associations, social scientists grew large enough in concert to become viable social actors. Developing potency, however, was only the initial step in a complex maneuver for social authority. Social scientists had also to develop a clientele for their advice and to educate the public in the desirability of expert reform. In the pursuit of these goals, the objectivity claims made by the recently unified professionals allowed them to define their role as organizers of everybody else. Claiming resistance to self-interest advanced these academics' own standing.

Some influential visionaries posited a neutrality independent of class location that performed several functions. It justified self-serving behavior, attempted by "social control" sociologists among others, by claiming authority as disinterested benevolence; because expertise transcended the self-interest of traditional class conflict, middle-class experts could counter both urban ethnic masses and plutocratic elites with power based in knowledge. This understanding also stifled social alternatives while legitimating existing arrangements. According to the historian Mary Furner, while objectivity "restricted open public advocacy of the sort that allied political science with reforms which threatened the status quo," it failed to "preclude administrative work or research that indirectly supported the interest already in power." Finally, the myth obscured the social scientists' role within—and dependency on—the subject of their inquiry: investigators methodologically removed themselves from the society, polity, or economy even though they themselves were neighbors, voters, and buyers. Political scientists, sociologists, and economists balanced reform with detachment in distinctive yet consistent ways.[27]

Political science suffered from multiple and competing agendas, despite the field's attraction to and attractiveness for governmental reformers. Three visions had support in the early period of professionalization. The Columbia School of Political Science, for example, claimed that through "the scholarly development of the discipline," it was training "professional public servants and equipping students to pass the civil service examination." This focus on administrative technique coexisted with a second desire: to transmit "values, knowledge, and patriotism as education for citizenship." The itch for involvement made actual government service a third option for political scientists. The first impulse proved the strongest in the decades between 1880 and 1920 as reformers hoped that apolitical managerialism could revitalize the possibility of political leadership. This paradoxical formulation—proposing to restore politics by taking it away from politicians—was to gain more and more adherents in the years up to the Depression.[28]

As of 1910, however, political science had not yet reconciled service with cerebration. Two historians of the discipline concluded that the political "science" of the first quarter-century was essentially "legalistic, descriptive, formalistic, conceptually barren, and largely devoid of what today [1967] would be called empirical data." And the news got worse. Because neither their province nor their approach was the subject of any consensus, political scientists struggled to define their field. In this search, admiring references to engineering were common, accompanied by untiring confidence in the discipline's future. A. Lawrence Lowell may be the most noteworthy of the optimists. As early as 1889, the New England textile heir had fittingly compared the study of politics to the examination of a carpet loom. Anyone attempting to study such a device "at rest, will find its mechanism hard to understand." Likewise, "the real mechanism of government can be understood only by examining it in action." By 1909, Lowell expressed disappointment at the distance between the professor and the practitioner. In so doing he implied that application validated knowledge: if engineers "paid little heed to the discoveries and opinions of applied science . . . would it not prove that those professors were off the track, or that their science was still in its infancy?" The envy of the social scientists could be no clearer: "But how much do statesmen turn to professors of political science?"[29]

In another camp entirely, hedging not at all, the scholar Woodrow Wilson dissented and held fast to the moralistic heritage of nineteenth-century clericalism. Despite his envied reputation in the field for his assertion, later embraced by Progressive reformers, that politics and administration should be separated, Wilson left no doubt as to his feelings about conceptions of the political scientist as political engineer. In an address of 1910 to a profession less than ten years old, Wilson demolished some of the APSA's most cherished pretexts and aspirations, looking to a bygone era when the clergyman represented both learning and virtue: "I do not mean that the statesman must have a body of experts at his elbow. He cannot have. There is no body of experts. There is no such thing as an expert in human relationships. . . . I do not like the term political science. Human relationships, whether in the family or in the state, in the counting-house or in the factory, are not in any proper sense the subject-matter of science. They are stuff of insight and sympathy and spiritual comprehension." His diatribe represented a rethinking of the statement he had made in 1901 that "leadership and expert organization have become imperative, and our practical sense, never daunted hitherto, must be applied to the task of developing them at once." Perhaps fittingly, Wilson's conduct of foreign policy almost exactly a decade after his APSA presidential address fell victim to the same moralism with which he clubbed his colleagues in 1910.[30]

Arthur Bentley attacked the young discipline from another angle. While Wilson and Lowell stood firm on the issue of political virtue, Bentley stressed the necessity of political scientists' commitment to the ideals of science. In his massive and at the time unrecognized landmark *The Process of Government*, he baldly and prophetically stated on the flyleaf that "this book is an attempt to fashion a tool." Striving to demythologize the terms of political analysis, Bentley envisioned a truly scientific study of politics without the intrusion of what he called "soul stuff." His view of the current state of the art was direct and unmistakable: "There is hardly anywhere a work on political science that does not, when it examines the phenomena of public opinion, either indulge in some wise and vague observation, or else make a frank admission of ignorance." Political science, as of 1910, clearly could please few of its own practitioners and even fewer of its constituents.[31]

If political scientists in the early part of the century were confused within the discipline and harassed by irreconcilable demands from without, sociologists faced an even worse situation. While politics constituted at least a readily definable field of inquiry, "society" proved to be broad and, more important, already claimed. In the words of one critic of the period, "Take from [sociology] what belongs to psychology, history, anthropology, ethics, civics, jurisprudence, economics, statistics, and charity administration, and there is nothing left of value." Upon receiving a questionnaire that asked members of the nascent discipline in the mid-1890s about sociology's relation to other fields, one anonymous respondent jestingly captured the essence of the situation. "The relation of Sociology to Political Economy, History, etc.," he wrote, "is *close*." The youngest social science, in terms of its professional group, was in its early years a splintered and besieged band of academics. Nevertheless, the promise of a science of society inspired great hopes within the ASS. In the ambitiously titled "The Meaning of Sociology," Albion Small laid out a program noteworthy for the grandiosity of its scientific claims and the fatuousness of its ill-defined terms: *"The final scientific problem is to ascertain the ratio of value for human experience of every factor which may enter into that experience."* The relationship of input to output of course defines mechanical efficiency, although Small refrained from the term in the 1908 article. A 1913 address, however, outlined his "Vision of Social Efficiency"; the mechanical metaphor persisted.[32]

Small's 1908 article contained a longer and even more obtuse claim about the value and methods of his field, no longer a discipline or profession but a "sociological movement," whose "common creed" could be reduced to two assertions. "First," he wrote, "the final judgment which men can pass upon anything of interest to men is discovery of its meaning in the light of all that can be ascertained about the whole process of human experience." Once that epistemological rabbit was pulled from the hat, Small continued, "all men should co-operate in finding out everything, and the relative value of everything, which is available for promoting the permanent interests of men." Such assertions, incredible to some even in an age when positivism was still a live option, invited cruel refutation.[33]

Writing from his vantage point in political science, Henry Jones Ford happily obliged, attacking Small's premises, methods, and aspirations in "The Pretensions of Sociology." He correctly identified Small's hyperbole about science as a smokescreen for extraordinarily foggy thinking on the matter and asserted that "sociology has not yet established any claim to be accepted as science"; no science could so clearly lack a definition of its proper subject. "It is clear that this movement, this faith, on its own showing, has no right to rank as a science or to set up any claim of authority," he wrote. Ford furthermore defended his home discipline, itself none too well established but with large designs on governmental influence: "If sociology lacks scientific validity, it cannot give safe guidance to any movement and its invasion of the political arena is an added peril." Lowering Small's "science" to the level of muckraking, Ford argued that "instead of inspiring caution, it encourages haste, levity, and sensationalism in dealing with social problems." The article concluded by attacking sociology on its moralistic habits of thought held over from a bygone century. Worse than being unscientific, "sociology commends itself to people who mistake reverie for thought and feeling for judgment; who reach emotional conclusions from sentimental assumptions."[34]

While Small stood far ahead of his colleagues in the audacity of his claims, he wrote from a position of prominence within sociology; his opinions carried weight. Scientific models of inquiry and control, or more properly the search for them, became common by 1910 as the profession attempted to fulfill the legacies of both Comte and Ward. The early practitioners of academic sociology pursued both professional and social dreams, inspired by the beneficence and efficacy they attributed to the extension of scientific knowledge to society. Some new professionals also announced that sociology would be the synthetic discipline organizing the findings of economics, political science, history, ethics, and philanthropy. Finally, these sociologists sought to *make* social progress; improvement no longer was presumed to be the result of automatic evolution. A vision in which science would discipline and enable humanity to control its existence mirrored the bright prospects of engineering. Promising scientific progress for all of society allowed sociologists to look out for themselves.[35]

Among economists of the early 1900s, the tendency toward scientism was muted, possibly by the still-unvanquished legacy of laissez-faire. Nevertheless, the economics profession was committed to a technocratic professional language (that of marginal utility analysis) and strict internal discipline through the purges of Richard T. Ely and Edward W. Bemis for "unprofessional" heresies. Orthodox viewpoints persisted longer in economics than in other fields. Possibly out of reaction against the "political" half of political economy, from which they recoiled, economists at once pursued social technique more diligently and less ambitiously than did their fellow academics. Instead of drawing overall plans for the direction of America as a whole, most reform-minded economists focused on the empirical dynamics of a single market, such as labor, railroads, or agriculture.

The first decade of the new century, then, found political scientists, sociologists, and, to a lesser degree, economists in search of realistically and not merely rhetorically scientific methods and aspirations. As a new professional type, the social scientist could claim a definite identity neither before the public nor within the various associations. In addition, despite frequent efforts to standardize disciplinary practice, eclectic thinkers—Veblen most centrally—continued to challenge the orthodox and inspire the restless. Engineering, by contrast, gained security and prestige from its longer professional history, its unarguably spectacular results, and its conscious reliance on science and practicality. Better-defined identities and cultural authority would later come to the social sciences, but in 1910 they still sought the earmarks of professionalism.[36]

PART TWO ·················

DEFINITIONS ·

···

1 9 1 1 – 1 9 1 8

That is the essence of scientific management, this

great mental revolution. . . . I know that perhaps it

sounds to you like nothing but bluff—like buncombe

—but I am going to try and make clear to you just

what this great mental revolution involves, for it

does involve an immense change in the minds and

attitudes of both sides.

FREDERICK W. TAYLOR TO SPECIAL

COMMITTEE OF THE HOUSE OF

REPRESENTATIVES, 1912

2

ENGINEERS AND EFFICIENCY

Amid intensifying public admiration of technical achievement, the cultural uses of applied science evolved between the late nineteenth century and World War I. The engineers' self-understanding and the social adaptation of technical terms, *efficiency* in particular, helped establish the credibility and prestige that later reformers envied. Engineering's rapid professional growth, theoretical innovations in personnel management, and inroads made by the Progressive sensibility in the engineering-managerial sector combined to create an ideological context favorable to advocates of applied logic and productive expertise. By World War I, engineering and society had become entwined in social theories, political languages, and new networks of technicians and reformers.

ENGINEERS AND SOCIETY

American industrial success derived from many sources: natural resources, internal and external markets, ingenuity, and a governmental climate conducive to commerce. Engineers sought to capitalize on their contribution to this success by organizing, publicizing, and rationalizing their profession. Highly attuned to salient issues of respect, reward, and responsibility, Herbert Hoover (1874–1964) helped to bring his colleagues the dignity he felt they deserved. Despite disclaiming that "the verdict of his fellow professionals is all the accolade" the engineer sought, Hoover gave considerable thought and effort to the task of improving engineering's image. In response to a playwright requesting technical details about mining, he wrote, in his own inimitable punctuation, "Ive whiled away many idle hours constructing a drama to represent to the world a new intellectual type from a literary or stage view—the modern intellectual engineer—theres more possibilites than you think."[1]

Hoover and many other leaders held that engineers' scientific training gave them a powerful commitment to objectivity, a regard for the facts that some of them took to be a matter of secular revelation. The president of the ASME stated in 1911 that "as we reverently discover and apply natural laws, we find new reasons and supports for . . . fundamental ethical conceptions." Their commitment to the facts led some engineers to think they could rationalize many aspects of the industrial order. In addition to positing that engineers could stand as disinterested buffers between labor and capital, Hoover sought specifically to attenuate the speculative side of mining by suggesting uniform practices of mine valuation. Scientifically informed professionalism, he believed, could rid the investment process of hucksterism and bring the engineer added respect as a counselor much like an attorney. In the mining field, Hoover developed a coherent managerial understanding notably more enlightened, especially in its stand on labor unions, than that developed by Frederick Taylor at roughly the same time.[2]

As its leaders stressed increasingly uniform college curricula and standards of practice between 1890 and 1910, engineering held fewer openings for those who, unable to afford higher education, rose from less skilled to more design-oriented positions; professionalism implied white collars, not dirty fingernails. Hoover endorsed this tendency, writing a college textbook that stayed in print until 1967. *Principles of Mining* (1909) covered the requisite technical practices and included extensive treatment of business matters like valuation and record keeping. The emphasis on professional conduct, addressed in a chapter entitled "The Character, Training, and Obligations of the Mining Engineering Profession," reveals Hoover's fundamental conception of his calling.[3]

After outlining the need for commercial savvy on the part of the aspiring engineer, the textbook turned to the benefits of the profession. "Every red-blooded man," he wrote, has a right "to be assured that his work . . . will build for him a position of dignity and consequence among his fellows." Not content to claim merely technical expertise, and unconsciously echoing Edward Bellamy, Hoover contended that "the very essence of the profession is that it calls upon its members to direct men. They are officers in the great industrial army." Still, the brightest days lay in the future, no matter how prosperous the present. While "the real engineer does not advertise himself," the "engineering profession generally rises yearly in dignity and importance as the rest of the world learns more of where the real brains of industrial progress are. The time will come when people will ask, not who paid for a thing, but who built it." This conception of industry closely follows Veblen's distinction between pecuniary and industrial occupations, a concept that influenced analytical political economy more widely after the war.[4]

In addition to writing the textbook and undertaking his activities within the profession, Hoover contributed a historical foundation to engineering's command of technique. With considerable help from his wife, Lou Henry Hoover, and a corps of translators, metallurgical chemists, and other experts, Hoover prepared a translation of the ancient mining text *De Re Metallica*. Originally written by the German physician Georg Bauer under the Latinized name Georgius Agricola and published posthumously in 1556, this vast source of mining lore, history, and technique was Western culture's first comprehensive study of mining and metallurgy. In financing the project from their private resources, the Hoovers did not only satisfy a love of history or old books. Their introduction to the translation, which appeared in 1912, claimed that "if the work serves to strengthen the traditions of one of the most important and least recognized of the world's professions, we shall be amply repaid." The handsome volume remains an impressive piece of scholarship and a classic in the history of science. In a sense, it marked the fitting conclusion to Hoover's career as a mining engineer: he had worked assiduously to give his profession the respectability of an ancient past, a rationalized and profitable present, and an autonomous future. Two years later he left engineering for good to enter political life.[5]

By the time of the release of *De Re Metallica*, the engineering profession had begun to confront the contradictions involved with being the offspring of both business and science. Heir to the respected legacies of cost effectiveness and practicality, the engineer could understand his social role in a number of ways. As objective applied scientists, some saw themselves as mediators between labor and capital; their allegiance was only to the facts. The same laws governing the physical world governed society, so discovery of these laws would lead to the possibility of rational social control, full employment, and economic stability. On the other hand, most engineers held a detached and even superior attitude toward the social realm, preferring to remain untainted by imprecise and politicized issues. Thus Frederick W. Taylor's landmark paper on scientific management had to be published outside ASME channels, because it considered ideas beyond the accepted limits of the profession. Even Taylor's position as a former president of the society did not enable him to get a hearing.[6]

Nevertheless, in the 1910s a minority of engineers sought to democratize the professional societies and disperse engineering's benefits to society at large. While this movement did reform the societies, especially the ASME, it must be stressed that a probusiness, antigovernment ideology dominated a generally apolitical and socially conservative profession. Paradoxically, this very distaste for politics uniquely qualified the engineer for political leadership in the eyes of some progressives, who reasoned that an aversion to politics removed the technician from the taint of corruption. Such logic took hold only among a small group of social outsiders, but the implicit approval of engineering as a model of government originated in the mythologized portrayal of the engineer in this the profession's heyday. A few engineers saw the incipient hero worship and tried to turn it to private advantage, while others—notably Morris L. Cooke and Herbert Hoover—understood their training as a social inheritance. Despite the small number of engineer-politicians, the point here is that the spirit of technocratic progressivism shared the premises of engineering: a supposition of objectivity, a stress on method, a belief in knowable and predictable laws, and a linkage of knowledge to control over one's world.

. . . .

A few engineers used their prestige to make overtly political inroads. Charles Steinmetz, for example, served as the president of the National Association of Corporation Schools in 1914. Here he began to graft his lifelong socialist inclinations onto the corporate ideal and to develop a social philosophy. The basic ideas introduced in addresses before the group came to a somewhat wider public as Steinmetz, never at a loss for words, began to write syndicated newspaper columns for the National Editorial Service in 1914. His literary output increased steadily: he placed articles in *Collier's* and *Ladies' Home Journal* and later published in *Harper's*, *American Magazine*, the *Survey*, and elsewhere. Also influential on a national scale was a small band of technical experts inclined toward social reform. In the years before 1915, Harrington Emerson, Morris L. Cooke, and, especially, Frederick W. Taylor based theories of social management on engineering principles.[7]

One of the most fascinating efficiency experts, Emerson largely created the field in tandem with Taylor. After studying mechanical engineering at the Royal Polytechnic in Munich, heading the modern language department at the University of Nebraska, and working as a researcher for the Burlington Railroad, he scouted an underwater cable route to Asia. In contrast to the doctrinaire Taylor, Emerson remained flexible in his definitions and commonsensical in his advice. "Efficiency," he wrote in 1909, "means the right thing is done in the right manner by the right man at the right place in the right time." Like Theodore Roosevelt, he drew on the old language of moralism and the new mythos of applied science: "Ideal, highest efficiency can be attained only through a combination of infinite goodness, infinite wisdom, and infinite power." Such inane definitions contributed to the dilution and vagueness characteristic of efficiency in the Progressive Era. Without making absurd claims, he nevertheless espoused a model of efficiency with far-reaching benefits. "True efficiency"—as opposed to "ideal, highest efficiency," one supposes—meant "ameliorated conditions for the worker, both individually and collectively—not only for the worker, but also for the employer—not only for the employer, but also for the corporation, and finally for the nation." For these gains to accrue, Americans needed to look to "engineering knowledge and practice" in many areas of life.[8]

Morris Cooke (1872–1960) continued to manifest some important intellectual currents of the age. His stress on knowledge as the primary element of social progress put him in close agreement with John Dewey's brand of pragmatism, which emphasized the role of "social intelligence" in the modern age. Cooke wrote in 1913 that "after scientific management has been introduced, the shop is run by the collective intelligence of the many." Also like Dewey, Cooke believed in education as an agency of social reformation, even if his attempt to make schools more efficient was poorly received in academia: the president of MIT wrote that Cooke's study of education read "as if the author received his training in a soap factory." Cooke expanded the domain of engineering from the study and control of materials and physical forces to the study and control of human beings. "Social engineering" is a literal translation of his definition of scientific management. "We are only beginning to teach," he wrote, "something as to the strength of men, the mechanics of men, the spirit of men; in a word, the economy of men. It is just beginning to dawn on us that there is a philosophy and an art and a science of human labor, with laws as definite as those of any other science." In the terms he used to understand himself, Cooke was a radical humanitarian. By supporting education of the masses, involvement of experts, and the consideration of society in mechanical, "objective" terms, Cooke and many other reformers sought to equip Progressive America for the modern age. Democracy would be improved, not outmoded, by their efforts.[9]

Cooke's paper "Some Factors in Municipal Engineering," delivered in 1914 to the ASME, illustrated the convergence of the engineering-managerial viewpoint with the reigning sentiment within public administration. The

Fig. 2-1. Morris L. Cooke (Franklin D. Roosevelt Library)

reformer reiterated the lessons he had learned in the Blankenburg administration and called once again for an enlightened professionalism. As a scientist, he argued, the engineer had a responsibility "for the development of
facts, regardless of whose advantage they serve." This commitment would
allow social rationality to triumph over self-interest, no matter how capably
represented: "I have in mind that the service of an engineer should be as the

service of a judge and as opposed to the service of a lawyer who confessedly seeks out and represents the interests of his client, and 'makes the worse appear the better cause.'" The engineering method, in its purified form, appeared once again as a force that could replace social contention with definitive answers.[10]

The discussion of Cooke's paper highlighted the tendency toward "scientific" political designs. One commentator reported that "city governments may be brought up to the standards which have prevailed for many years in England and in Germany by taking advantage of the skill and ability of the engineer." Another participant, Robert B. Wolf, saw the engineer to be "destined to work out the great social problems of the world as well as the industrial problems." Conventional connotations would be transformed: "The word 'politics' must be made to have a new meaning and the duty of the engineering profession is to make it synonymous with the highest kind of idealistic service." Herbert Hoover would echo this message repeatedly over the next decade.[11]

Wolf's reasoning for this conclusion revealed an ever more popular logic: "In the very nature of things [the engineer's] idealism is practical. His grasp of material facts and laws insures an idealism which is workable and for this reason progressive in its accomplishment of social and political reforms." The engineer, "above all others, is qualified to solve the great vital problems of our municipalities." Note the parallels with pragmatic philosophy. Society values engineers because their idealism works and because they deal with observable forces of cause and effect. The engineering method—a "grasp of material facts and laws"—insures predictability and consistency in an uncertain world; raw idealism could promise no such reassurance. Finally, societies are the product of "the forces of nature" and not tradition, irrational beliefs, or dumb luck. Cooke, Wolf, and the other municipal engineering advocates, in short, understood society in the same terms used by proponents of "scientific" sociology and political science.[12]

Throughout this period, Cooke involved himself in projects aimed at strengthening efficiency and democracy simultaneously. Serving in the Taylor Society leadership, he maintained that "Taylor is the only industrial technician who has proposed anything approximating an adequate philosophy for industry." Working in a city government, Cooke closely monitored the movement for municipal reform using expert scholars and administrators. Challenging the ASME leadership, he persisted in trying to implement more democratic membership practices. For Cooke, the connections between engineering, administration, and politics were clear. "To me," he proclaimed, engineering is "'the art of organizing and directing men and utilizing the forces and materials of nature for the benefit of man.'" Because politics is "'the science of government,'" Cooke concluded, "our profession cannot evade not only an active participation in it—but I am tempted to say a dominant position in it." This idealized understanding of engineering colored Cooke's entire political economy; the iron logic of the "one best way" undermined his pleas for democracy. Cooke's biographer maintains that despite "his good will, the doctrine amounted to paternalism."[13]

Amidst all of the reform enthusiasm within engineering, Frederick Winslow Taylor (1856–1915), the man Cooke later called Master, continued to demarcate the broadest expanse of ideological landscape. Taylor perpetually sought control; his unstable personality made him try to impose order on all aspects of his life. The one best way could be determined for any activity, from shoveling coal to drilling steel to playing tennis with an illegal racket. Not content with mere profitability, Taylor tried to develop objective measures of efficiency more accurate than capitalist gains in the market. The laissez-faire model of the economy held no sway, for Taylor promised that the world of work and production could be, and must be, rationally managed for universal benefit. According to one historian of managerialism, Taylor approached any problem in industry or in life "not as a 'mystery,' skill in which could only be acquired by years of use and wont, but as a definite logical structure of cause and effect which could and should be mastered along the lines followed by a scientist dealing with a new gas or liquid." The neutrality of the technician could overcome class, ethnic, and other antagonisms. Such a conception, broader in implication than the business efficiencies proposed by Emerson or by Taylor's associates Frank and Lillian Gilbreth and Henry L. Gantt, promised to alleviate the most pressing social problems worrying the Progressive middle class. Even Taylor's phrasing, his call for a "complete mental revolution," appealed to this powerful and emerging constituency: ideas could reshape society in fundamental ways, thus discharging the building economic tension tapped by anarchists, Wobblies, and other "radicals."[14]

Just as efficiency as a standard of success promised to make profit obsolete, managerialism could also transcend class-defined social conflict. Although Taylor wanted nonpartisan science to replace existing relations of opposition and oppression in an industrial nation, that neutrality was illusory: he and his kind would engineer the reconciliation in terms of *their* rationality, not some unarguable truth. His bold claims found an enthusiastic audience, especially after 1910, as Taylor invoked an ill-defined public in-

Fig. 2-2. Frederick Winslow Taylor (S. C. Williams Library, Stevens Institute of Technology, Hoboken, N.J.)

terest wholly compatible with the common weal being upheld by the muck-raking press. The profoundly undemocratic nature of Taylor's system, however, led him to commit a mistake later repeated by Herbert Hoover. While looking to public opinion for support, partially through effective image making, both trusted administrative technique to the exclusion of public advice;

neither man listened very carefully to the citizenry he wooed. The vehemence with which Taylor defended his work from criticism gives one indication of his unshakable confidence. Distinguished by its exceptional neuroses, inflexibility, and singleness of vision, his personality inspired others who saw in the historical context demons similar to those chasing Taylor. A zealot whose time had come, he initiated his own "mental revolution," one carried on by his followers.

In 1909, Taylor began giving annual lectures at the recently founded Harvard Business School, whose dean, Edwin Gay, saw the Taylor system as a major advance in management and made it the focal point of the first-year course.[15] Two years later, Taylor published *The Principles of Scientific Management*[16] and helped found the Society to Promote the Science of Management, which was later renamed for him after his death. Even in the movement's most successful era, however, few firms could afford to implement Taylor's entire system. Enormously multiplied paperwork and labor hostility to the speedup imposed by stopwatches and to changing rates for piecework muddied in real life a theory so coherent in the abstract. Nonetheless, Taylor became famous as a recognized authority whose advice was widely sought. This stature resulted not from overwhelming implementation of the system by industry, but from a railroad rate hearing involving one of America's greatest jurists.[17]

Like Taylor, Louis Brandeis possessed a personality extraordinary enough to hold in tension some striking contradictions. Jewish and raised in Louisville, he moved easily within the Brahmin class of Boston society. He earned over $2 million in legal fees between 1878 and 1916, yet he carried a reputation as "the people's attorney." While pioneering the use of sociological jurisprudence, he still served with distinction as a Supreme Court justice with vision to see beyond the reform fad of the moment. It was Brandeis who catapulted efficiency into the public imagination. For all of Taylor's intensity and the suave graciousness for which Emerson was known, neither could match Brandeis for brilliance or charisma. In the meetings leading up to his appearance as counsel for the Traffic Committee of the Trade Organizations of the Atlantic Seaboard at the Interstate Commerce Commission rate hearings of late 1910, he insisted that all of his witnesses agree on the name of what they advocated. Merely to do so showed daring: Taylor openly disliked Emerson and would soon damn Frank Gilbreth as a "fakir." The consultants proposed names like Taylor System, Functional Management, Shop Management, and Efficiency, all distinctly lacking in panache. At that point, as

Brandeis later wrote, "it seemed to me that the only term which would properly describe the movement and also appeal to the imagination, was 'Scientific Management,' and as I recall it all present were ultimately unanimous in the adoption of that term." Unlike the managerial crusaders, Brandeis knew the value of marketing a clever phrase.[18]

The hearings themselves assumed the character of an ongoing bureaucratic grind until Brandeis made his famous charge: the railroads did not deserve rate increases because their massive inefficiencies could be remedied with projected savings of $300 million annually. Reporters quickly turned the claim into the "million dollars a day" catch phrase that focused attention on the case and moved it to the front page. The railroads disputed the figures on waste—"All Scoff at Brandeis" read the headlines—and dared him to name his own salary, come to work for them, and back up his accusations. He called the bluff and seized the moment to score an even more audacious public-image coup. He decided to devote all his time on the case in the pursuit of public interest; in the words of the *New York Times* headline of November 30, 1910, "Brandeis to Teach Roads without Pay—Spurns Proffered Big Fee—Is Now Representing Shippers for Nothing because He Does Not Care to Be Too Rich." Because primitive systems of record keeping prevented the railroads from answering inquiries about their own operations, the attack on the shippers stalled. As the *New York Evening Post* reported, "The railway cross-examination may be said to have begun in a spirit of flippancy and ended in rather awkward silence." At the close of the hearings Brandeis filed a ninety-four-page brief, nearly half of it devoted to claims by the efficiency experts, that convinced the Interstate Commerce Commission to deny the rate increase. Despite their bitterly divided ideas, Emerson, Gilbreth, and Taylor benefited enormously from the exposure.[19]

The case at least temporarily transformed traditional political economy. Despising managerial speedups even more than they distrusted the railroad barons, railroad workers came out against the shippers. Instead of mediating between labor and capital, the engineering viewpoint had united them against the managers and efficiency consultants. Brandeis encountered loud opposition from both working-class and plutocratic spokesmen, but the growing power of middle-class managerialism still made him, and the efficiency experts, public figures of high repute. The perceived solution to class war, as Hoover and others had asserted, was for the objective engineering-managerial mind of the middle class to come to the fore. Efficiency presented an all-win situation; society played a zero-sum game no longer. Governance therefore became an issue of technique, a problem to be solved. The rate hearings helped to mobilize even further a growing movement to rationalize not only technology and business but politics and people.[20]

EFFICIENCY: VARIATIONS ON A THEME

Support for the application of expertise to society quickly became apparent in the press, as roughly one hundred magazine articles in the years 1911–14, most prepared by self-proclaimed experts, proposed to define and encourage efficiency. The craze, which reached a peak of popularity and bizarreness in 1914, attracted the usual charlatans and hangers-on found in the midst of any popular enthusiasm. Efficiency groups sprouted like clover, as did all manner of efficiency publications. Some merely restated the gospel of success or clothed old bromides in the new jargon of efficiency. (See figs. 2-3 to 2-5.) Luther H. Gulick, for example, wrote *The Efficient Life*, a vastly mixed assortment of advice on sleep, attitude, and other concerns. (A noted YMCA leader, Gulick had co-invented the game of basketball with his student James Naismith.) "The first step in the cure of constipation is to get into the right frame of mind," Gulick asserted, before advocating as a more extreme measure "a ride upon a hard-trotting horse." Because "many city business men in middle life have bodies that disgrace them," he invoked the ideal of Theodore Roosevelt, "who sometimes leads the simple life, who often leads the strenuous life, but who always leads the efficient life." Roosevelt, healthy living, and moral condemnation ("disgrace") appeared frequently in the efficiency discourse.[21]

One article entitled "More Brains, Less Sweat" demonstrated the mechanism, the popularity, and the stern moralism of the movement. "The idea of half doing a job in double time; the spectacle of people foozling and fuddling, without plan, without standards; the whole idea of wasted labor and wasted material" were anathema to these efficiency advocates. At this stage of the development of technocratic progressivism, the evil of waste still touched a highly sensitive nerve attuned to conceptions of work dating at least to Benjamin Franklin. Elements of the theory, however, had been substantially updated to address workers in the age of Homestead and Pullman, as the article showed. "There is something infectious about these 'efficiency engineers,'" it said. (Note how physiological and engineering metaphors coexisted; *vitality* and *efficiency* could mean much the same thing.) "They don't talk grandly about the dignity of labor, and then waste it. But they give

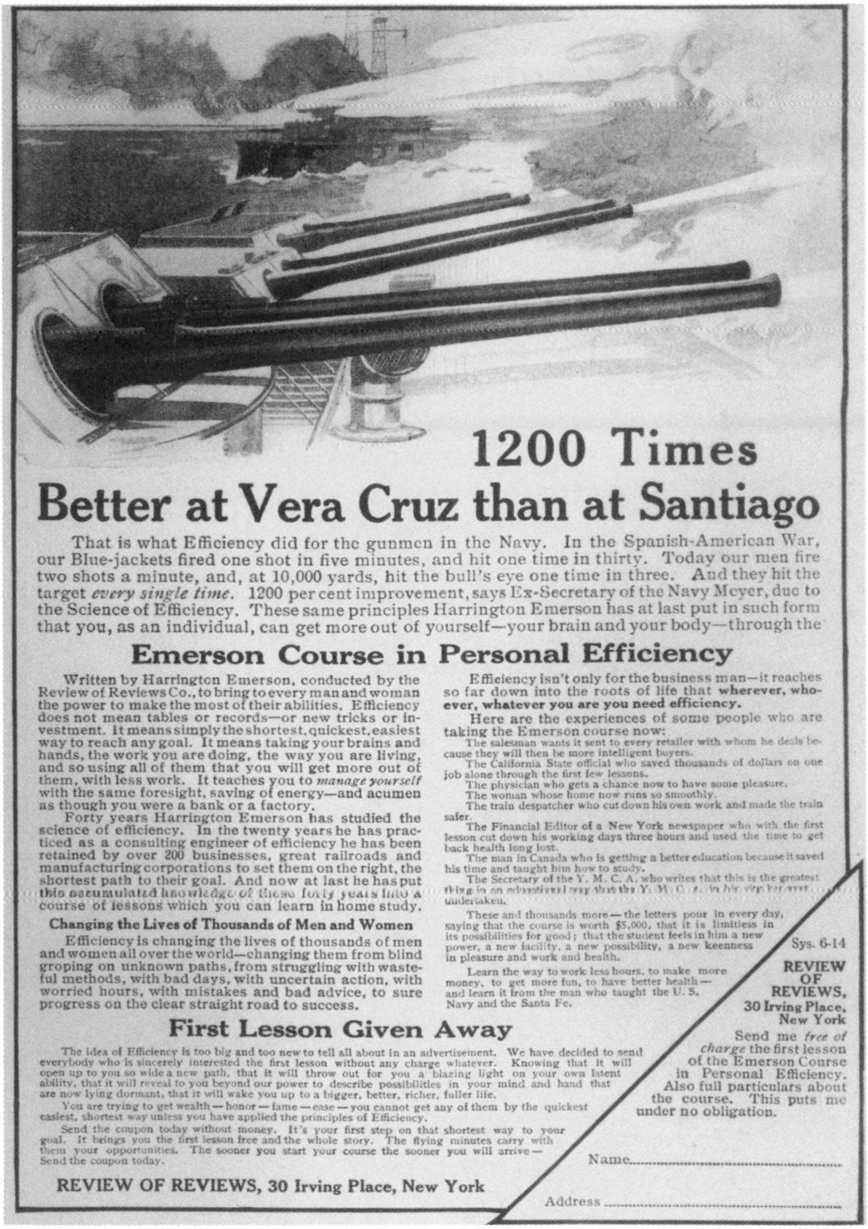

Fig. 2-3. Advertisement for Harrington Emerson's course in personal efficiency, 1914

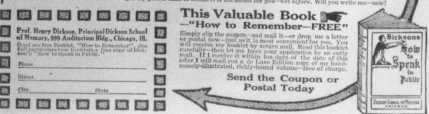

Fig. 2-4. Advertisement for Dickson School of Memory showing resemblance of efficiency hucksterism to older appeals

dignity to labor by using labor carefully." The theme of control, common within the idealized understanding of engineering, figured prominently: "Where formerly there had been a machine running a man, there was a man running a machine." Such pronouncements, squarely in the midst of the efficiency literature, would be augmented and transformed in the later work of that writer, the same Walter Lippmann who had quit socialism in Schenectady.[22]

The vague and malleable definition of efficiency allowed various renderings in lay periodicals. The *Independent* magazine, for example, featured an "efficiency service," in which Edward Earle Purinton would answer "any question that may be asked in relation to personal efficiency, health, work, and business." Purinton wrote that "efficiency begins with wanting something so hard the whole world can't stop you," and that "only efficiency conquers fate." In addition to inspiring his reader with platitudes, Purinton used the technical connotations of his topic to group humanity with levers and dynamos. "Man is the only machine we have never learned how to use," he wrote, metaphorically tying himself to both idealized engineering and, possibly unconsciously, philosophic pragmatism. (See fig. 2-7.) The conclusion to the article fused its twin emphases on moralism and mechanism: "Part of you is spirit—part of you machine. Listen to the spirit—then grip the machine!" For the benefit of readers who wanted to do so immediately, Purinton included a thirty-question "Personal Efficiency Test." The prospective machine gripper answered the following queries, among others, with percentages:

> Do you know where your greatest power lies?
> Do you drink three pints of pure water daily?
> Are you independent, fearless, positive?
> Do you enjoy art, music, literature, and the presence of little children?
> Is all your clothing made loose, to allow blood and nerves free play?

Further articles were promised on topics such as Cash and Efficiency and Home and Efficiency.[23]

In the same issue of the magazine, "ten efficient men" defined the term in vague and confusing ways. Charles W. Eliot, at the time president emeritus of Harvard, called "will-power the tap-root of efficiency," while the department store owner John Wanamaker preferred "the word 'service' to 'efficiency.'" Judge Elbert Gary of US Steel opined that "the key to optimism is altruism" and sounded more like a social gospel preacher than an industrial giant: "Not only is the crushing of competitors wasteful and harmful, but ac-

Fig. 2-5. Once again, the origins of efficiency language appear in hygiene, success, and dietary appeals

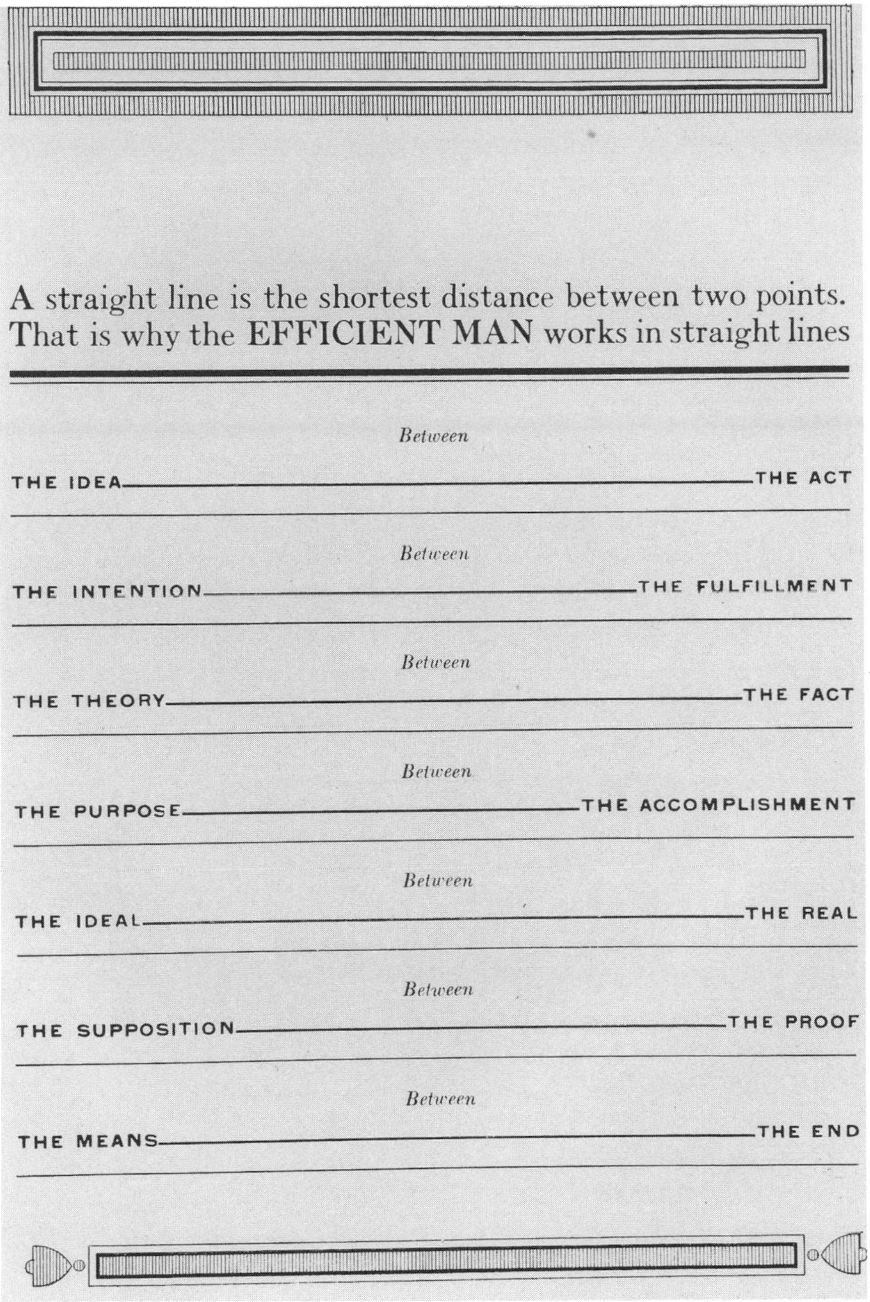

A straight line is the shortest distance between two points.
That is why the **EFFICIENT MAN** works in straight lines

Between

THE IDEA————————————————————————————THE ACT

Between

THE INTENTION————————————————————THE FULFILLMENT

Between

THE THEORY————————————————————————THE FACT

Between

THE PURPOSE——————————————————THE ACCOMPLISHMENT

Between

THE IDEAL————————————————————————THE REAL

Between

THE SUPPOSITION————————————————————THE PROOF

Between

THE MEANS————————————————————————THE END

Fig. 2-6. Diagram from System *magazine, 1912*

SYSTEM

A monthly magazine devoted to the improvement of business method—
two dollars the year—twenty-five cents the copy—enterered as
second class mail matter at Muskegon, Michigan—contents copyrighted

Volume IV   MAY 1903   Number I

Get the thing done. The tag ends of unfinished business are time-consumers. They drag on. They multiply. They take ten minutes to do, if they are done to-day; two hours, if they are done to-morrow.

Get the thing done. Keep your eye off the clock. Keep your interest undivided. The new problem will be the more easily tackled when the old one is out of the way.

Get the thing done. That is system. System stands at the door and denies admittance to every interrupting detail. System sees that every facility is at hand—at the finger's end. System keeps things away from you until you are ready for them.

Create your system as you go along. When, by no fault of yours, a thing goes wrong, it is a symptom that there is a lack of system. Sit down then and there and devise a system which will insure you that that particular thing will never again go wrong. Don't wait till to-morrow to devise the system. Get the thing done.

There is satisfaction and success in a finished article. There is danger and delay in even an unfinished detail. Proceed calmly, forcefully, quickly, but not hurriedly. Get the thing done.

Fig. 2-7. Editorial, System *magazine, 1903. This editorial anticipates several later developments: pragmatism's emphasis on results, the deletion of data not fitting one's model, and the absence of personal responsibility— the method is credited with accomplishment.*

tual friendship may continuously be applied to competitive business." Gulick, by now a celebrity in the movement, was the most explicit of the lot as he hoped for a society where engineering could improve democracy: "Can you develop the geniuses who will parallel in the social world what our inventors have done in the world of steam and electricity?" The crusading mayor of New York, John Purroy Mitchel, testified to the benefits of municipal reforms such as rationalized budgeting and the direction of policemen, social workers, and educators toward new urban purposes. Louis Brandeis insisted that "efficiency is the hope of democracy."[24]

Theodore Roosevelt, never too comfortable away from public life, was interviewed on efficiency in business by *System* magazine in 1913. The article praised Roosevelt with terms similar to those used by Gulick several years earlier. Not only did Roosevelt advocate efficiency, he lived it: "Judged by either his personal or his executive achievements, one cannot escape the conviction that Mr. Roosevelt is the most efficient human machine of our time." For his part, Roosevelt affirmed once again that his wilderness movement lay at the foundation of the current excitement. In addition, he argued that Taylorism and its variants possessed great national utility. Technique, he implied, could replace previous bonds of civic obligation: "We couldn't ask more, from a patriotic motive, than scientific management gives from a selfish one." Such bald formulations were rare, but increasing numbers of reformers aiming to update outmoded concepts of politics shared technocratic sentiments.[25]

. . . .

While the popular attention to various representations of efficiency came and went, the managers who spawned the movement struck out into territory beyond that claimed by Taylor. With the election of Harlow Person as president in 1913, the Society to Promote the Science of Management began to broaden its membership base and range of interests. The contrast between Person and Carl Barth, Taylor's most faithful and rigidly orthodox follower, illustrates the nature of the split in the organization. Barth steadfastly refrained from any social speculation, focusing solely on machines. He did share a moment of poetry with the United States Commission on Industrial Relations by stating that "my dream is that the time will come when every drill press will be speeded just so, and every planer, every lathe the world over will be harmonized just like musical pitches are the same all over the world." His specialty was drill presses, for which he developed a famous slide rule that calculated speed, feed rate, and other parameters for various applications. Person, the dean of the Tuck School of Business Management at Dartmouth, symbolized the managerial impulse within Taylorism. Before Taylor died, Barth was his favored disciple, as Henry Gantt and Frank Gilbreth espoused decreasingly orthodox formulations of the founder's doctrine. Under Person's direction, the society moved decisively into managerial matters in the 1910s and 1920s.[26]

Since 1912, the Taylorites had been joined by another group, simply known as the Efficiency Society, which claimed members from occupations more diverse than business and engineering. The society's leadership included Gulick, Harrington Emerson, Edwin Gay, and three prominent social scientists: Irving Fisher, E. A. Ross, and E. R. A. Seligman. From the ranks of government came the British social welfare advocate Albert Beveridge, the juvenile court innovator Ben Lindsey, and the municipal reformers Frederick Cleveland and Henry Bruère of the Bureau of Municipal Research in New York. Journalists who joined included Oswald Garrison Villard of the *Saturday Evening Post*, Walter Hines Page of *World's Work*, and Adolph Ochs of the *New York Times*. As might be expected from the diversity of its membership, the Efficiency Society preached a message of unity: the various camps were all indebted to "apostles of a single faith—a faith so large, so universal, that it benefits all fields, and in varying guises inhabits all nature, animate and inanimate." Note that the presumption behind all of the moral and social grandeur is still that of the one best way, which so obsessed Taylor; logic dictated that a scientifically validated efficiency would supply answers to the perennial ambiguities of human society. Another contributor to the Efficiency Society journal wrote that "whether it be in the running of a household, an industry, or municipality and nation, there is a certain line of thinking common to all, which, if followed through, is bound to bring about the best results in each." His simplistic positivism testifies in part to the optimism of the moment.[27]

The search for formalized techniques within various areas of life can also be seen in a growing emphasis on business management. Consider, for example, the history of *Engineering Magazine*. In 1910, its nineteenth year of publication, the periodical carried articles about boilers, generators, and the like. Six years later the title was changed to *Industrial Management*, and technological articles gave way to an emphasis on managerial issues such as personnel selection, record keeping, and employee motivation. Similarly, within the Taylor Society, Henry L. Gantt advocated advanced forms of cost accounting as necessary steps in the measurement of efficiency. He also called it "very unfortunate that the term 'efficiency engineer' was ever invented," preferring instead that "a more correct appellation would be 'management' or 'industrial' engineers." Some managerialists thus merged claims made on the credibility of their two forerunners, engineering and big business, as they participated in a wider debate over social utility.[28]

While some technocratic reformers moved their emphasis from the physical world to the industrial firm, others outlined principles of political administration. One professional leader suggested that the engineer possessed "knowledge of mechanism, supplemented by his knowledge of the inner as well as the physical man." The profession was therefore "destined" to bring forward "this human element as the controlling factor in determining the major policies of management of our lives, our industries, our nation." He concluded by asking his audience to "hope that each [engineer] may receive the best possible remuneration to the end that he and his family may have the largest possible share in the good things that make for comfortable and happy homes." Self-interest, professional expertise, and public aspiration continued to create a confusing range of possibilities for American engineers.[29]

Along similar lines, Henry Bruère openly borrowed from engineering as he proposed an approach to urban criminology. He commended engineering as home to a well-defined technique and acknowledged the conclusion reached by Cooke and others that "engineers have alone [among professionals] found a permanent vocation in municipal work." Bruère then painstakingly defined the engineering method, which "consists of applying scientifically determined knowledge to the execution of a particular problem, and the use of ordered and analyzed facts as a basis for formulating conclusions in respect of that problem." He concluded that "repeated application of the engineering method" could establish a legitimate "technique" for guaranteeing results.[30]

The article raises several pertinent questions. Why is a public administrator writing in a mechanical engineering journal? Where are legislators, constituents, and citizens in this model? Why is engineering so appealing in its contribution to a technique of administration? How does efficiency bear upon the process of police protection? From his post at the BMR, Bruère had access to a staff of political scientists, accountants, and other figures well acquainted with urban government. That he so explicitly invoked engineering as the model for municipal practice indicates the appeal of engineering far beyond technical practice.

In this hectic and promising time for technomanagerial reform, Frederick Taylor enjoyed a few years of fame and vindication before his death in 1915.

The blaring tributes at a well-attended memorial service celebrated his transition to an odd sort of ethicotechnical sainthood. Quoting a French priest, Morris Cooke declared that "the love of God is the Taylor system of the inner life" before continuing in his own words: "In every part of the world . . . were those who called him Master." The presiding minister's benediction eulogized a great humanitarian who attempted, "by readjustment in the chaos of life's jumbled parts, to set things right"; Taylor's "lifelong purpose was to save men's souls, to bring them into possession of themselves and thus assist them . . . to serve their fellows and glorify their God." John Mapes Dodge, at whose Link-Belt plant Taylor had refined conveyer belt production, stressed in his tribute the religious connection even more explicitly, linking Taylor's paternalism to a utopian faith in human possibility. "Many others have prayed for an industrial social millennium, expecting it to come from spiritual grace through lapse of time," he noted, setting up his straw man carefully. "But Dr. Taylor [the degree was honorary] not only saw the possibilities of the future, but he did more." Based upon "eternal truth," Taylor's work "told in detail exactly how this long-hoped-for condition might actually be accomplished at once."[31]

The memorial service merely condensed ongoing efforts to connect engineering and religiosity in these years, especially within Taylorism. Henry L. Gantt had written that once the "world is controlled by deeds rather than words"—by engineering rather than politics—Christianity would change "from a weekly intellectual diversion to a daily practical reality." The most daring of these analogies were drawn by Cooke. As well as calling Christianity "the Taylor system of the inner life," he made an even more audacious claim a decade before Bruce Barton popularized a similar comparison: "You will have to go back a long way to find out who established the Taylor System. A long while ago an obscure carpenter, of a despised race, set himself up to teach his people a new system of doing business. . . . All that Frederick Winslow Taylor, one of the greatest engineers who ever lived, did in his life time of effort was to translate into a practical, profitable, working formula, the Sermon on the Mount." The intermingling of technical and moral justifications for rationalized social relations powerfully illustrated the crossroads to which political and ethical language had come.[32]

In their hyperbole, Cooke and Dodge articulated the cultural forces woven together in the Taylor myth. It portrayed efficiency as an all-American Christian virtue, advocated the improvement upon old notions of spiritual truth by the engineering method, and exhibited the unbounded hopefulness characteristic of a believer in human perfectibility. Scientific management promised all of this—"at once," no less—and at no cost; both labor and capital, once enlightened, would benefit from the new gospel. Taylor himself had repeatedly boasted that his comprehensive system was no mere speed-up, gimmick, or panacea. While claiming benefits for all of society, however, Taylor kept a close focus on the industrial firm and despised efficiency "cranks" like Purinton. Ironically, Taylorism caught on much less tenaciously with business leaders than with Progressive Era reformers in search of a technique and an inspiration for the management of social change.[33]

. . . .

In the aftermath of Taylor's death, Charles P. Steinmetz (1865–1923) became perhaps the most celebrated and opinionated engineer in America. The fascination with expertise in this period encouraged Steinmetz's generous estimate of his own opinions and served to make him a public authority on almost anything—conservation, religion, education, and especially economics and politics. William Randolph Hearst's *American Magazine* began one article in April 1919 by asking,

> "You don't think we will have Bolshevism in *this* country, do you?"
> That question is being asked by men and women all over the United States. It is the greatest question before the nation to-day. And because it is, *The American Magazine* wanted it answered by someone who can speak with authority.
> Probably no man in America is better fitted to do this than Charles P. Steinmetz, who is regarded by many as the greatest electrical engineering genius in the world.[34]

While Steinmetz adhered to an idiosyncratic socialism, his credentials made him palatable; once again, technique negated ideology. To reassure their readers that this was no bomb-throwing agitator, biographers and magazine writers repeatedly disavowed any revolutionary intentions on his part. The *American Magazine*'s disclaimer was typical: "In theory, he is a Socialist. . . . But he is so sane and broad-minded in his views that he antagonizes neither the progressive employer nor the intelligent worker." Steinmetz closely followed the progress of the Bolshevik revolution; he aided a drive to buy tractors for Russia, sat on the advisory board of *Soviet Russia Pictorial* magazine, and wrote to Lenin offering to help with the electrification of the young nation. Lenin refused the offer but responded with an autographed photo that according to several sources hung conspicuously on the wall of Steinmetz's GE laboratory.[35]

Fig. 2-8. Charles Proteus Steinmetz (General Electric Archives)

His one nontechnical book, *America and the New Epoch* (1916) held that humanity had passed from the age of individualistic competition to the age of collective cooperation. Steinmetz repeated this statement countless times throughout his public and private writing, basing all of his social ideas on one historical generalization. What great sea change forced such a shift in social understandings? The productive capacity of modern industry had out-run consumption; firms competed not for resources but for markets. Eventually, *"the price, forced by free competition, is below the cost of production."* The resulting economic instability caused unemployment and poor management of resources. Since unemployment bred social conflict, cooperative economic reorganization would bring social harmony. This impulse toward what came to be called industrial democracy tied Steinmetz more closely to management reformers, many of whom rejected Taylor's more autocratic methods after 1915.[36]

While he argued that "the purpose of industrial production is the welfare of the members of society, the producers," Steinmetz allowed that industrial mass production had changed the role of the worker. Calling for unemployment insurance, old age insurance, and disability insurance to alleviate the "three great fears" of the working man, Steinmetz also understood that industrial labor failed to satisfy the human creative impulse. He accordingly supported a four-hour workday, with increased efficiency and decreased consumption of inessentials making the extended leisure possible. He considered the era's abnormal increase of consumption "to a large extent artificial and unnatural, fostered by the producers." He continued: "A considerable part of the world's work of today is not production, but is advertising, selling, and [attempts] to increase the production by stimulating demand where it did not exist." In a logical world, production of essentials during short, managed workdays would meet society's basic needs and allow ample time for the personal fulfillment that work provided in some previous age: "Leisure will stimulate educational interests in every conceivable direction, and man will become a highly informed and much more intelligent and self-expressive creature." The possibility that increased leisure might only encourage more consumption of inessentials never entered his argument.[37]

Who would "control production to correspond with the legitimate demands for the product, cease all production for mere profit, and [dismantle] all organization for the purpose of creating a demand where it does not exist"? According to Steinmetz, the government should assume this function, taking as its model the corporation: "All that is necessary is to extend methods of economic efficiency from the individual industrial corporation to the national organism as a whole." Because "in our nation it would require not merely that the political government take over industrial control," a new form of administrative organization was necessary. In Steinmetz's view the industrial corporation had proven its managerial superiority over elected government. Despite Veblen's critique of businessmen's pecuniary mindset, Steinmetz thought business would *become* government and run the state more efficiently. Such a prospect "presupposes a powerful, centralized government of competent men, remaining continuously in office." Administrative continuity precluded frequent changes of leadership (mandated by an inexpert public) so that "in the national Government by the cooperative organization of corporations, there would be no election of officers." Instead, talent would be self-selecting; one would rise through the ranks.[38]

Steinmetz included only a minimal democratic component in the new state: the tribuniciate. To align efficiency with democracy, he needed some medium for electoral participation that would not interfere with the effective management of the state. Functioning as a sort of emergency brake, the tribuniciate was "an inhibitory power outside of the industrial government; a power not organized for constructive administration and executive work, . . . but invested with an absolute veto to stop any action of the industrial senate which is against the public's interest." Such a two-tiered arrangement would preserve democratic participation on larger social questions while allowing skilled administrators the necessary continuity and flexibility to create the good society. Nowhere in *New Epoch* did Steinmetz discuss psychological or spiritual benefits of the industrial corporate state. His deeply held faith in a rationalized state took precedence over the consideration of more intuitive human needs.[39]

By basing a system of corporate altruism on a life of rare intelligence and an instance of extraordinary employer tolerance—he was allowed to work at home and at a riverfront cottage—Steinmetz mistakenly attributed the same generosity and flexibility to all corporate organization of capital. In the first two decades of the century, especially when laissez-faire failed the demands the Great War placed on its weak coordinating capacities, advocates of centralized responsibility saw better government as more efficient government. Efficiency found its highest expression in the corporation, so therefore the corporation should become the government. Seeing only the facile logic of such reasoning and not its failure to define terms, Steinmetz did not realize the many uses to which efficiency's fluid and imprecise definitions lent themselves.

This standard appealed to Steinmetz because of its empirical basis. While most social questions involved ambiguous value judgments, waste was an obvious evil, economy an unchallenged virtue. The book's opening paragraphs revealed his aggressive confidence: "The following does not represent my sentiments, but gives the conclusions drawn from historical facts which of necessity follow from the preceding causes, regardless of whether we like them or dislike them. . . . We must entirely set aside our sentiments and wishes, and, like in any physical or engineering problem, draw the conclusions which follow from the premises, whether they are agreeable or not." Unlike Henry Adams, who searched in vain for a science of history, Steinmetz confidently endorsed a naive positivism; his powerful method could be applied directly to social issues in order to solve them "like any engineering problem." (He included in *New Epoch* an account of the efficiency to be gained by adopting the 20-hour workweek and the 200-workday year; such an arrangement would yield 4,675 leisure hours a year for a promised efficiency of 53.4 percent.) Politics and ideology could surely be tamed by scientific methods, or so he reasoned.[40]

The argument for scientific historical causation in the preceding passage belies a need to have the world make sense in the terms to which Steinmetz was accustomed. Nevertheless, subjective issues still troubled him. Despite the objective and quantitative advantages of a conventional definition of efficiency, he had yet to include the more psychic aspects of existence. So Steinmetz stretched the word further: industrial efficiency "is the efficiency of man as a cog in the industrial machine, but not his efficiency as a human being. *What then, in our purpose, is efficiency? It is to make the most of our lives* and our industrial productivity is but a part of a means to that end, although it is not the end." Steinmetz thus discounted the prevailing systems of both industrial organization and industrial reform. He retained efficiency as a social cure-all but moved the locus of its pursuit from the factory shop floor to the corporate boardroom.[41]

In the years just before the war, a renegade from the Taylor ranks came to a similar understanding. Henry L. Gantt (1861–1919) concluded a life of management reform by positing that it was the captains of industry and politics who constricted America's productive capacity; laborers were merely foot soldiers in what Veblen had called the pecuniary drive for profit. The war brought the issue to a point of crisis, and Gantt responded with a blunt call for a state organized on engineering principles, one similar to that proposed by Veblen. Gantt's writing before 1916 had been concerned with personnel training, record keeping, and other technical aspects of industrial organization. After Gantt read some of Veblen's work, the tenor of his writing

Fig. 2-9. Henry Laurence Gantt (S. C. Williams Library, Stevens Institute of Technology, Hoboken, N.J.)

changed markedly. Beginning with the paper "Efficiency and Democracy" delivered to the ASME in 1918 and culminating in *Organizing for Work* (1919), Gantt challenged conventional understandings of democratic politics with a conception of a new productive order. He also took practical steps to implement his ideas and, in doing so, likely inspired the Technocracy movement of the next generation.[42]

After finding, like many others, "that our political system alone was not adequate to the tasks before it" in modern warfare, Gantt pointed accusingly at businessmen accustomed to "harvesting dollars," not producing goods. America's need for manufacturing skills, not salesmanship, made many of the dollar-a-year men aware of the real leader of industry—the engineer, not the financier. The peacetime future was to be marked by a continuation of wartime productive leadership, in both industry and politics. To do this, "opinions must give way to facts, and words to deeds, and the engineer, who is a man of few opinions and many facts, few words and many deeds, should be accorded the leadership . . . in our economic system." Politics would be adjusted accordingly. "Real democracy consists of the organization of human affairs in harmony with natural laws" and included an end to what Gantt called "debating society methods." Instead, America needed "not more laws but more facts."[43]

In addition to his writing, Gantt supplied much of the motive force behind the formation of a vanguard of politicized engineers—more than two years before Veblen proposed such a thing. At the ASME convention in December 1916, Gantt helped organize rump meetings of fifty people, not all engineers, who formed a group to "increase the purchasing power of a day's work in New York City," the site of the convention. They called themselves the New Machine, but little else happened. A long, windy letter was sent to President Woodrow Wilson asking him to invest the machine with the authority of presidential empowerment, but he never responded and the group soon disappeared. Nevertheless, in a newspaper interview published in the month before his death in 1919, Gantt remained unwavering in his conviction and made an apocalyptic prediction: "I have no faith in governmental machinery as government is at present organized. . . . The system cannot last. It is breaking down under its own weight."[44]

Gantt and his accomplices staked out odd but definite claims to historical footnotes. According to Gantt's biographer, the first Russian five-year plan was entirely plotted on Gantt charts. Another New Machine member, Walter Polakov, studied electric power issues in the war period, wrote Steinmetz's obituary for the *Nation*, and in 1925 published *Man and His Affairs from the*

Engineering Point of View, an intriguing assortment of philosophical and technological arguments. He went to Russia and returned to the United States in time to witness the Technocracy movement before working in the New Deal. The "chief theorist and general manager" of the machine, paid out of Gantt's pocket, was Charles Ferguson, formerly a "special agent of the United States Department of Commerce." His social theory was, if anything, even more radical than that of Gantt, to whom he served as something of a guru.[45]

Ferguson's books *The Great News* and *The Religion of Democracy* never caught the public imagination. His writing, in tone and content utopian and hortatory, found probably its widest audience with the publication in the *Forum* of "The Men of 1916," a *Looking Backward* pastiche set in 1967. The men of 1916 were, of course, the engineers, who seized the moment and re-organized democracy along logical and technological lines. On his death-bed, the old engineer, who had been there, tells how the great moment had come. "In the striving for the subdual [*sic*] of brute forces to the uses of the human spirit," Ferguson's hero sermonized, "there is no place for any authority that is not intrinsic and self-vindicating—*the authority of the man that produces the real goods*." Thus "the triumph of art and engineering is the achievement of civil liberty."[46]

The old system, with its "slathering sentimentalities and sneaking sub-terfuges," fell in tatters before the phalanx of technologists. "I can hear the yawp," recalled the old man, "of the college boys in Broadway: 'America first—three cheers for the engineers!'" The assertive vanguard "didn't wait for the crowd to say what it wanted. Crowds never say. We assumed that the crowd wanted to be led by men who knew the way. We knew the way." As in *Looking Backward*, logic precluded argument at the appointed hour of the historic transition: "'Consent of the governed'—of course. We expected everybody to consent to that kind of government. Everybody did." War had lit the spark for the movement, the old man recalled, and the men of 1916 carried the day. The time when engineers were asked for advice "was the hour for our *tour de force*." In contrast to feeling underappreciated, "We were full of pride and confidence and had begun to feel that the machine-age was somehow peculiarly *our* age, that the lawyers and the clergy had had their turn and that our inning had come." Unfortunately for the engineers, such was not quite the actual scenario.[47]

Ferguson and other engineering visionaries used the preparedness debate to discuss America's course of social action. Like *efficiency*, *preparedness* was an imprecise and elastic term used by commentators of many ideological persuasions. Theodore Roosevelt, still a powerful Progressive symbol and an advocate of both modern managerialism and old-fashioned blood and glory, joined lesser-known figures in a fierce debate with noninterventionists, many of whom feared a plutocratic lust for wartime profiteering. While the backdrop of mechanized warfare brought a special urgency to issues of industrial reform, it also forced preparedness advocates to differentiate themselves from a most unpopular paradigm of social efficiency.[48]

For academics educated in German universities, scientists awed by German discoveries, and managers seeking German standards of productivity, complex issues surrounded the European war. For many of these Americans, mobilization had little to do with hatred of Huns, and the gradual involvement of the United States on the side of Great Britain was by no means uncontested. An idealized perspective on scientific Germany informed the good society sought by these men. Once Germans became villains, preparedness advocates like C. E. Knoeppel were deluged with hate mail. Squirming, Knoeppel tried both to equivocate and to stand fast, first denying his opponents any moral ground and then insisting on performance as the primary standard of judgment: "Regardless of whether our sympathies are with the Germans or the Allies, regardless of whether the Germans are right or wrong in this war, and without considering the methods they use to achieve their ends, all must admit that they are showing the rest of the world an efficiency both as to organization and methods that is nothing short of marvelous." Americans could learn from Germany to pursue the "elimination of politics from things influencing the welfare of the people" and "intelligent direction through expert guidance."[49]

Industrial experts took part in most of the preparedness arguments, many of which included postwar expansion of the engineer's social role. In 1916 a White House press release stressed that "preparedness, to be sound and complete, must be solidly based on science," and many engineers eagerly maneuvered their profession into the potentially vital work. After America had "an efficient mechanism for carrying on the work" of fighting the war, a larger issue emerged. "The point is this," wrote Knoeppel. "Will we plan now, to make this mechanism permanent, or will we let it rust and fall apart after the war?" As of 1915, Knoeppel and the Taylorites hovered on the fringes of the business center. With the onset of war, their promises of increased productivity merged with national self-interest, and engineering enthusiasts received a wider hearing.[50]

At the same time, some managerialists moved closer to labor than to capital; the evils of inefficiency, as Gantt and others had pointed out, indicted less the workers than the plutocrats removed from manufacturing processes. Thus the tendency toward industrial democracy caught on, and organized labor did make significant gains during the war. The reformers leading the way in this direction included Ordway Tead, a Taylorite who wrote for the *New Republic*, and Henry Dennison, the New England office aids manufacturer who served in the war mobilization and headed the Taylor Society from 1919 to 1921. In 1919 the society sided with workers in the bituminous coal strike and favored an eight-hour day for steel strikers. The main impetus for this conciliation, in marked contrast to Taylor's bitter hatred of unions as havens for institutionalized loafing, was the appearance in 1915 of the reasonably generous study of Taylorism by Veblen's former student Robert Hoxie. While not hypercritical, it did raise questions about the methods and goals of modern management, and many Taylorites grudgingly admitted that unions might in fact serve a purpose. In keeping with their new mainstream image, reformist engineers favored moderate Gompers-style unionism over the more radical strains of the Industrial Workers of the World. Many Taylorites argued that because unions would undoubtedly get stronger, forward-looking managers should begin rapprochement as soon as possible. The relative absence of self-righteous managerialists after Taylor's death hastened this conciliation.[51]

Before and during the war, the Taylor Society underwent a series of internal debates that reflected its increasingly varied constituencies and still-conflicting goals. The unflagging sense of destiny remained; these reformers concurred with Steinmetz that a new epoch awaited. Engineering, they claimed, could transform the most powerful cultural systems, political, social, and philosophical. Polakov, for example, told a reporter that after the elimination "of *all* friction in her industrial system," America could reduce the workday.

> "With production simplified and power utilized to its fullest capacity, we could probably produce all we want in much less than six hours; and with distribution simplified we would have no trouble in securing the product for our own enjoyment."
> "Socialism?" I asked.
> "Engineering," he corrected.[52]

War fever inflamed such grandiosity. The Taylor Society met in December 1917, and the discussants' positions on the question of wartime organization reveal intense division of sentiment. Henry Kendall, a bandage manufacturer, argued that crisis necessitated centralization. Using the metaphor so

close at hand in the period, he asserted that "a perfect organization that constitutes a single machine, whether for fighting or for some other purpose, is
what we must have." Such a machine must be "organized at the top" and
should be "an organization of experts with complete responsibility." Disagreement with this view came quickly. One Taylorite took issue with
Kendall, citing Germany once again. "No machine will ever win the war,"
he argued. "Germany gives us the finest example of the mechanistic idea,
fully worked out. . . . Is not America's mission to take the measure of that
machine and prove its weakness?" Morris Cooke, meanwhile, claimed that
organization from the bottom up, not the top down, merged efficiency and democracy.[53]

As of World War I, engineering and efficiency had contributed significantly to the creation of a technicopolitical ethos in which technique, moralism,
and self-interest jostled for primacy. Especially as cultural relativism represented a troubling prospect to those uncomfortable with multiple sets of values, the promise of a unified morality excited many Americans. Goodness
and virtue, tainted by Victorian connotations, no longer served as social im-
peratives in an industrial nation. In some measure, *efficient* functioned as a
scientific-sounding all-purpose favorable adjective. At the same time, the
possibility of definitive standards of morality, determined by supposedly objective means, attracted many progressives who sought to remedy the ills of
the time while still retaining a belief in human perfectibility. By endorsing
the efficiency ideal, many Americans could adhere to a "muscular Christianity," which retained an affinity with traditional Protestantism while still
appearing unquestionably modern.

This quasi-religious fervor began with the Taylorites' attempts to relegate
both capital and labor to roles subordinate to the middle-class engineer-
manager. A certain degree of proselytizing proved necessary to convince
these groups of the manager's primacy. Following the lead of Frederick Taylor himself, the cadre of administrator-engineers promoted an industrial ca-
suistry; the Taylorites invested in efficiency a substantial sum of Christian
moral capital. Descriptions of the movement contain repeated sacerdotal
overtones; "converts," "pilgrims," "orthodoxy," and "heresy" describe various aspects of the effort to teach American industry the One True Faith,
which had been "born" at Bethlehem Steel. Taylor grew furious with imita-
tors and could only recommend four "disciples" to give efficiency advice:
Cooke, Gantt, Barth, and Horace K. Hathaway. Other pretenders were
deemed "cranks and charlatans"—not heretics—by Taylor's admiring biographer.

....

What did advocates of social efficiency gain by adopting engineering ideas? The word *politics* connoted the shortcomings of civil debate as its vocabulary and premises confronted imposing aggregations of private power, not the state power feared and restrained by the inventors of the American republic. In the aftermath of the Gilded Age, efficiency promised a nonpolitics of administrative competence, virtuous in the face of existing corruption. Traditional Protestant/republican moralism—stressing frugality, honesty, public-mindedness, and respect for individual rights—served to undergird twentieth-century conceptions of political efficiency. Political language expanded to account for new perceived realities while revising some comfortably reassuring ethical bases.[54]

In an era when the term *progressive* connoted a steady, teleological, restrained pace of improvement, *efficiency* implied change while at the same time suggesting security. The smoothly humming social machine envisioned by these reformers promised harmonious eradication of social problems; a bloodless revolution, as predicted in *Looking Backward*, would remake the world. The Efficiency Society and the Taylorites suggested an end to discord by focusing on the consensus of rational agreement, and they made efficiency at once reassuringly static and demonstrably effective through images of balance. Charles Buxton Going, the editor of *Engineering Magazine*, wrote that "efficiency does not demand or even encourage strenuousness. It does not impose or even countenance parsimony. It merely demands equivalence—equivalence between power supplied and work performed." This peculiarly American paradox of kinetic change made stable appears to have contributed to the ubiquity of efficiency claims in this era.[55]

Beneath the inflated rhetoric and frequent silliness, the quest for the one best way joined Taylorite ideas with later managerial and political theories. Especially in the work of Frank Gilbreth, an associate of Taylor purged for overemphasizing time-and-motion studies, the one best way marked the initial step in a comprehensive revision of the workplace. In contrast to their artisanal forebears, industrial workers were told by the scientific manager what that one way was to be. As a consequence, the worker lost autonomy, and this separation of conception from execution helps to define the modern workplace. Marx's theory of a worker-capitalist split was thus complicated by what some have called the new class, as white-collar managers mediated between laborers and stockholders. In addition, the premise that rationalized processes were beyond contention appealed to many managers in an age when labor problems often brought gunfire. Discussing the Taylorites in comparison to municipal reformers, the historian Dwight Waldo wrote that "since 'best ways' rely on facts, they are, of course, True, and not proper subjects for differences of opinion." Scientific method could thereby contribute to social "progress" by delimiting the field for cultural difference. By focusing on means, furthermore, efficiency advocates also curtailed debate over ends. The preoccupation with technique, with best methods as opposed to best purposes, marked this particular pursuit of social progress. A similar grammar quickly became apparent in other reform communities, especially in response to the trauma of wartime.[56]

3

STRUCTURING A NEW REPUBLIC

In the years that engineers professionalized and efficiency became a fad, intellectuals and reformers closer to traditional politics than to business or technology sought to redesign the state. In New York, such institutions as the Russell Sage and Rockefeller Foundations, the Carnegie Corporation, the *New Republic* editorial offices, and the BMR comprised a vanguard of the future in which academics and other experts devised and implemented plans for managed social change. The foremost intellectuals of this world—Herbert Croly, Walter Lippmann, and John Dewey—began to think and write using apparently scientific concepts. Influential but often unaccountable political networks also formed. Academic social scientists backed away from moral reform movements while seeking to strengthen the ethos and methods—scientism and quantification, respectively—that would dominate the field after the war. In addition, World War I forced Americans to be more conscious of the promise of science and the demands of industrial civilization. In the context of these developments, rational social management became an ever more promising option.

THE *NEW REPUBLIC* INTELLECTUALS

As laissez-faire political economists slipped in influence early in the twentieth century, new public bywords, civic goals, and theoretical investigations could enter American political economy. In a setting in which many middle-class Americans relished new challenges yet needed reassurance, Herbert Croly (1869–1930) published *The Promise of American Life* in 1909 and provided intellectual leadership to reformers seeking to remake government along managerial lines. It should be stressed that some intellectuals re-mained sensitive to the pitfalls of pure technique and searched for guiding national ideals. Indeed, the tensions between democracy and administrative modernization in the work of Croly and Dewey clearly demonstrate the dilemma in which many reformers found themselves. Nevertheless, Taylor's "complete mental revolution" and Croly's plans for a technique of reform shared a significant debt to idealized engineering.

While Croly was an unlikely and often uncomfortable advocate of rational reform, his past did hold continuities that led him to such a stance. After being baptized—physically—into Comte's religion of humanity, Croly grew to adulthood under the steady influence of his father's Comtean thought. "Society," wrote David Croly to his son in 1887, "is an organism, controlled by laws of development which when discovered can be modified by man himself." This proved to be a concise statement of the way many political reformers would view engineering: knowledge of universal laws, discovered by scientific methods, led to rational control. His father's Comtean posi-tivism; a Harvard education under William James, George Santayana, and Josiah Royce; and the atmosphere of reform in the early twentieth century made Croly the heir to several diverse and mutually energizing intellectual traditions.[1]

In *Promise*, Croly argued that America had grown enormously large and complex since originating as a mercantile and agricultural nation. To avoid the excesses of monarchy, the early model of American government had specified a weak executive, dispersing power first to the states and, under Jefferson's influence, to the yeomanry. Anticipating Walter Lippmann, Croly characterized Jeffersonianism as "drift." But the late nineteenth century showed the weak executive model to be archaic, because the multiplication of industrial size and complexity necessitated centralized, responsive gov-ernment on a scale commensurate with that of the great corporations. Big-ness was simply a fact of life in the industrial age, and America needed a government strong enough to act as an effectual participant in national af-fairs. Croly wanted no autocrat but appealed to a broad-based democratic movement that would retain Jeffersonian ends—democracy—through more modern means. The best people to run the state, those inspired by the scien-tific spirit of excellence, should be allowed to rise on the basis of their talent to govern. In short, America needed an active government capable of effica-cy on the national level. The era "demands a distribution of economic power and responsibility which will enable men of exceptional ability an excep-tional opportunity," Croly wrote. One such man was Theodore Roosevelt.[2]

Croly's ideas came to the fore in the extraordinary presidential campaign of 1912 when Roosevelt used *The Promise of American Life* as the centerpiece of his New Nationalism. He did so while trying to appear more progressive than Woodrow Wilson, steer the Republican party away from the influence of William Howard Taft, and offer a mainstream alternative to socialists leaning toward Eugene Debs. Like Wilson's and Louis Brandeis's New Freedom, the New Nationalism sought greater governmental efficiency, and both confronted the issue of corporate domination. While Brandeis sought to pare down bigness so as to assure relatively fair competition not far removed from a laissez-faire model, Croly saw a rationalized, centralized state to be the way out of the economic warfare he feared Brandeis endorsed. Still, the similarities between the two programs were borne out in the years after the election. Croly came to gain Wilson's trust and respect, and Croly's *New Republic* led the support for Brandeis's appointment to the Supreme Court. The Brandeis who so actively praised the Taylor system and the Croly who advocated centralized authority of experts in governmental affairs both contributed to the stature of rational reform.

In *Progressive Democracy* (1914), Croly urged his audience to embrace scientific principles of administration, but not without "a new faith, upon the rock of which may be built a better structure of individual and social life." In his brotherly tone, Croly defined "live-and-help-live" as the post-laissez-faire motto of the "progressive democratic faith," which was "a spiritual expression of the mystical unity of human nature." Still, the themes drawn in *The Promise of American Life*, sometimes even more boldly delineated, coexisted with the soul stuff. The authority of an expert administration, Croly contended, "will depend on its ability to apply scientific knowledge to the realization of social purposes." Because of the similarities between scientific management's role in business and the way general administrative staff functioned in government, the logical conclusion was clear: "The successful conduct of both public and private business is becoming more and more a matter of expert administration." Furthermore, Croly understood science to be capable of providing a method from the bottom up, as it were, whereby democracy could be made to work: the competence of the polity, enhanced by progressive education, would make possible an industrial democracy and a truly progressive social order.[3]

Also in 1914, Croly helped found the *New Republic*. Under the influence of his coeditor Walter Lippmann, the voice of humanistic reason within Croly's books was muted as an emphasis on administrators, technicians, and methods distinguished the magazine. After World War I, Croly would return to and accentuate the fraternal dimension of his social thought, but by that time the reform climate had chilled to the point where neither Croly nor his ideas carried nearly as much influence as they had in the Progressive period. In the interim, however, Croly's journal of politics, arts, and opinion established itself as an innovative and respected venture. Financed by the effectively unlimited wealth of Willard and Dorothy Straight, the magazine proved to speak for and to many intellectuals and politicians. No longer did reformism revolve around direct democracy and moral uplift or muckraking and the protection of women and child workers. Instead, Croly, Walter Weyl, and Lippmann advocated a partially technocratic progressivism that was pragmatic in its hardpan moral tone and in its stress on cause-and-effect reasoning. They envisioned a new republic national in scale, culturally modern in aesthetics, and domestic in political orientation. A strong executive, able to lead and administer the American leviathan, needed to be supported by experts in all phases of government.

A great blooming of American arts and letters, beginning in Greenwich Village, only blocks from the magazine's offices on West Twenty-first Street, seemed to validate Croly's and the Straights' faith in an intellectual and artistic renaissance. With painting and literature in the lead, the arts appeared to be lending an aesthetic dimension to the administrative and economic promises of American life. Many of these artistic and literary voices found expression in the *New Republic*. In the magazine's first twenty-five issues alone, the roster of contributors still inspires a certain awe. Dewey, Santayana, and Royce represented philosophy, Charles Beard and Ray Stannard Baker carried the banner of reform, and Robert Frost, Amy Lowell, and Van Wyck Brooks addressed literature and poetry. Within the first year of publication, H. G. Wells, Ford Madox Ford, Roscoe Pound, Lewis Mumford, Theodore Dreiser, A. O. Lovejoy, Ralph Barton Perry, Conrad Aiken, and Harold Laski, among others, came on board in some capacity. Croly possessed a keen eye for talent, enjoyed a fat budget, and designed a handsome format for his magazine.[4]

While Croly deserves substantial credit for the magazine's excellence, his coeditors did much to offset his lumbering prose and other weaknesses. Walter Lippmann began an illustrious editorial career at the *New Republic*, and his contribution to the progressive ethos still stands as an impressively coherent and reflective body of work. Walter Weyl, in contrast, was near the end of his life and did not leave an intellectual monument on the scale of Croly's or Lippmann's. He nonetheless held his own with the quiet, "yogi-like" Croly and the intense, brilliant Lippmann. A Ph.D. in economics and extensive research into railroad matters gave Weyl an ease with statistics that he combined with a powerful empathy. According to Lippmann, Weyl could justify progressivism "by the statistics of the social facts as well as by moral denunciation." Alvin Johnson, who left the *New Republic* to head the New School for Social Research, recalled that Weyl "looked like a saint and fundamentally was one."[5]

The New Democracy Weyl sketched in his book of that name closely resembled Croly's social vision. Even more than his colleague, however, Weyl was attracted by the possibilities that engineering inspired. His fondness for industrial metaphors allowed him to sustain an argument more vividly than the plodding Croly even as they reveal an underlying mechanism. Joining the advocacy of a more responsive and active state, Weyl claimed that the current government "answers to the needs of the people . . . ineffectually, like a clumsy, ancient machine which utilizes only one or two per cent of the power applied to it." His invocation of efficiency, broadly defined as always, did not merely connote typical civil service reforms. The call for a capable "engineer" to design and operate the new model is clear: "All this [civil service] efficiency is important, but a still greater efficiency on a far higher plane is necessary if we are to democratize our industrial and political life." Weyl's failing health limited the time he spent on reform politics, and he died at the age of forty-six in 1919.[6]

Much more than Weyl's, the contributions of Walter Lippmann (1889–1974) to the dialogue over the shape of the new America proved to be important and sustained. From the beginning of his literary production in 1913 with *A Preface to Politics* to his advice to Lyndon Johnson on Vietnam, Lippmann spanned, shaped, and scrutinized "the American Century." His books of 1913 and 1914, *Preface* and *Drift and Mastery*, outline a new American state both akin to Croly's and distinctly original. His later attacks on rational reform stand out sharply against this background.

At the age of twenty-four, Lippmann set out to record his intellectual influences. The result, *A Preface to Politics*, was urgent, serious, and often graceful. While his former Harvard Socialist comrade John Reed called him one "Who builds a world, and leaves out all the fun," Lippmann was an enormously talented writer—and one with a mission. Reed's poem continued: "He sits in silence, as one who has said; / 'I waste not living words among the dead!'" Despite *Preface*'s name-dropping and leaden gravity, the book still persuades. Its emphasis on the irrational nature of humanity, inspired by Lippmann's loyalty to the British social theorist Graham Wallas, partially obscures a cogent and sustained justification for technocratic politics. Holding that man shapes his world, that technique matters more than ideals, and that science offers useful lessons to the statesman, the *Preface* fits into the quest for a technique of rational reform. Like his colleague Croly, Lippmann regarded Theodore Roosevelt with favor, calling him the "working model for a possible American statesman at the beginning of the Twentieth Century."[7]

Lippmann rejected the idea of "a static government machine" and had no patience with the "tinkering reformer." He wrote later in the book that "government is not a machine running on straight tracks to a desired goal." Instead, "a machine must be run by men for human uses." The concept of control, of what he would shortly call mastery, endured as one of Lippmann's guiding principles throughout his life. Thus, "the type of statesman we must oppose to the routineer is one who regards all social organization as an instrument. . . . Call this man a political inventor." Such images of invention and control applied precisely to engineering. Lippmann specified, on several occasions, ideals squarely in line with the goals of applied science. "The object of democracy," he wrote, was "to harness political power to the nation's need." In light of the scale and spirit of engineering triumph, he stressed that "it is absolutely essential that men regard themselves as moulders of their environment." Lippmann later praised the transcendent technician as an example for the new order: "We shall be making our own house for our own needs, cities to suit ourselves, and we shall believe ourselves capable of moving mountains, as engineers do, when mountains stand in their way."[8]

In *Preface*, Lippmann sought to move beyond the limits of his muckraking and radical pasts. Specific political programs, he believed, would come and go; rhetoric was cheap. Instead, he sought to infuse a spirit and method of inventive reform into the American body politic. The new spirit would derive from the technician's sense of mastery, a nearly mystical concept. Man's "domineering impulses," in this view, "find satisfaction in conquering things, in subjecting brute forces to human purposes." This will to power, Lippmann predicted, would be "the social myth that will inspire our reconstructions." Along with myth, America needed method, because "method matters more than any particular reform." Thus inspired by the possibility of control and undeceived by cosmetic social proposals, Americans could make a modern state overcome routine and stasis. Science and invention animated Lippmann's visionary project, one in which "there will be a premium on inventiveness, on the ingenuity to devise and plan. There will be much less use for lawyers and a great deal more for scientists." Later in this ode to technique, Lippmann praised two groups among whom Taylorism held great sway: "industrial organizers" and engineers. The same man who wrote "More Brains, Less Sweat" in 1911 apparently still believed in a more sophisticated version of the efficiency message.[9]

Lippmann avoided many of the youthful excesses of *A Preface to Politics* as he wrote *Drift and Mastery*. The sobriety of tone, fascination with method, and sensitivity of insight remain, but a substantial originality marks the second book. Name dropping and unevenness gave way to a clarity of purpose as Lippmann sustained a tightly wrought argument for the primacy of science in the pursuit of "mastery." With an apprenticeship under Lincoln Steffens and a three-month stint with the Lunn administration in Schenectady long behind him, Lippmann moved to a new intellectual position removed from both muckraking and socialism. In so doing, he became technocratic progressivism's poet laureate. Seeing through the vagueness of the efficiency ideal, calling it "a word which covers a multitude of confusions," Lippmann worked instead with the vocabulary of Veblenian anthropology, pragmatism, and scientific managerialism.[10]

Veblen's insistence that a culture's advance is led by its technology anticipated Lippmann's contention that industrial and economic factors "have played havoc with the old political economy." Lippmann noted too that "engineers . . . have something more than a desire to accumulate and outshine their neighbors." He also invoked an "instinct of workmanship," calling it a cultural factor able to "temper the primal desire to have and to hold and to conquer." For both men, the state of humanity's tools and technique, its technic, determined its degree of civilization.[11]

In a clever logical maneuver, Lippmann shifted from an appreciation of pragmatism to an approval of technical expertise. He agreed with the pragmatic stress on the primacy of experience and its invalidation of tradition and equated "the modern intellect" with "this habit of judging rules by their results instead of by their sources." Problems arise with the complexity of the modern world, however, and "to act for results instead of in response to authority requires a readiness of thought that no one can achieve at all times. You cannot question everything radically at every moment." As a result, the technical expert comes to humanity's aid. "I have to follow the orders of my physician. We all of us have to follow the lead of specialists," he proclaimed. "We cannot be absolute pragmatists." Lippmann thus substituted the authority of expertise for that of tradition: "Where we have to accept dogmas without question we do so not because we have any special awe of them, but because we know that we are too ignorant, or too busy, to analyze them through." In this idealized transition, authority became demythologized, stripped of its antiquated ornamentation by the solvent of science.[12]

Lippmann sprinkled references to the possibilities of the scientific method and spirit throughout *Drift and Mastery*. In so doing, he articulated a broad social program that included a sophisticated version of social engineering; knowledge and mastery, science and technique, expertise and power are closely aligned in this scheme. While "we have felt reality bend to our purposes," the future appeared brighter yet. A growing body of opinion "looks to the infusion of scientific method, the careful application of administrative technique . . . lured by a future which we think is possible." For Lippmann, science was not an abstract piety; its value lay in its connection to action. It "is the culture under which people can live forward in the midst of complexity, and treat life not as something given but as something to be shaped." If the point that knowledge leads to control were not sufficiently clear, he reiterated: "In our world only those will conquer who can understand."[13]

. . . .

In the pages of the *New Republic*, Croly, Weyl, and Lippmann helped to define and advocate the methods by which America could become a modern state. Articulating a managerial and scientific, and less conventionally moralistic, language of social efficiency, the magazine issued a challenge to progressives. Few attacks on prostitution, confusing ballots, or monopolies appeared in the journal. Instead, the prospect of an expertly managed and rationalized state, home to cultural and artistic diversity, guided the magazine. As the European war continued to encroach onto America, the journal refined the call for centralized administrative power and advocated science as the model for modern liberal thought. An unsigned editorial published during the magazine's first year elaborated on the post-Progressive, protoliberal ideal: "Genuine opinion is neither cold, logical judgment nor irrational feeling. It is scientific hypothesis, to be tested and revised as experience widens." Such an idea was already "playing havoc with the old crusted folkways" by 1915. Even before publication began, an unsigned and undated memorandum presumably from Croly had proposed that the magazine "would stand, consequently, above all for a higher quality of human expression in American life—for moral freedom, intellectual integrity, social sympathy, and improved technical methods in all practical and fine arts." These goals quickly came into conflict, and the fixation on technique frequently overwhelmed more tender-minded objectives.[14]

The magazine's hopes for a science of reform dovetailed with the managerial "mental revolution" promised by the late Frederick W. Taylor. Croly recruited several Taylorite management reformers to contribute: Morris Cooke wrote a number of unsigned pieces, and Ordway Tead, a former Socialist, was identified by his initials. One editorial looked forward to the day when "the personnel executive will be paid by the three parties to industry—the organized employers, the organized workers and the government." Labor unions, still opposed to management reformers before the war, presented "a serious menace to productive efficiency in America." Society, the writer argued, could not afford to have the "major contribution" made by Taylor and others "sacrificed" to the failure of labor organizers and managers to come to agreement. Industrial productivity, the precursor to true abundance, thus figured prominently in the political economy of the magazine.[15]

While industrial management provided broad possibilities for change, the greatest opportunities and challenges lay within political administration. The *New Republic* commended Morris Cooke's application of engineering expertise to Philadelphia politics, while Mayor John Purroy Mitchel had given New York "a municipal instead of a political machine" by unseating the Tammany candidate. Structural change became imperative: because American political institutions were based on English precedents and on abstract political ideals characteristic of the late eighteenth century, the "American democracy soon outgrew its institutional equipment." The magazine declared that "our greatest need in America to-day is a working agreement between democracy and science."[16]

A science of politics and administration would be crucial to fill that need. One unsigned editorial posited that "the business of politics has become too complex to be left to the pretentious misunderstandings of the benevolent amateur." Elsewhere, Charles Beard predicted that "political science is to be the greatest of all sciences. Politics and physics are to be united, but the former is to be the bondsman." John Dewey, a frequent contributor, criticized the "preoccupation with lofty principles logically arranged" and bemoaned the current state of the discipline. The work of observing social changes, of forecasting consequences, and of controlling their impact "gets poorly done," he wrote. "Social control becomes a matter of luck." Irrationality and superstition, the antitheses of the modern temper, had no place in the emerging liberal ideology.[17]

Because of its commitment to productive performance in both industry and government, the *New Republic* frequently expressed dissatisfaction with existing administrative arrangements, especially the bumbling conduct of America's preparation for the war. In 1916, Croly contributed an article on preparedness to the *Annals of the American Academy of Political and Social Science*. Events would both vindicate and nullify his argument, for while hostilities did necessitate preparation, Croly was falling into the age-old trap of fighting the last war. Echoing Theodore Roosevelt, Croly asserted that "the American nation needs the tonic of a serious moral adventure." The prospect of such a tonic proved a bracing dose and made Croly contradict his long-standing belief in rational planning: "We shall have to take the risk of preparing first and of deciding later just what we are preparing for." The impetus for such a stance was less an ambitious internationalism than domestic reform; the threat of war challenged America "to use more foresight, more intelligence, and more purpose in the management of its affairs." American politics, grown sluggish and ineffectual, would be jolted awake; "an army and a navy large enough to be dangerous may introduce into American domestic life a useful ferment—one which may prove hostile to the prevalent spirit of complacent irresponsibility." Croly and others who hoped for a domestic political rebirth from the war found instead that it could stall progressive momentum, but memories of the Spanish-American War still invigorated.[18]

The *New Republic* joined other advocates of early militarization with its criticisms of inefficiency. A series of articles throughout 1917 and 1918 uncovered paralyzing bottlenecks in the administration and procurement of ships, coal, food, munitions, and manpower. Always in the background of wartime preparedness lurked the domestic implication; Croly's strong executive, informed by experts, continued to personify the new politics. "Efficient government which may have looked academic in happy days of surpluses is now becoming a matter of life and death," began one editorial of 1918. Calling the current arrangements "bulging and improvised war machinery," another unsigned editorial criticized the legislative body bent on denying the executive branch the requisite authority: "Many Congressmen believe government to be an automatic machine which can be perfectly designed in advance on paper, and then operated by a continuous flow of money." But this war could not be fought by a nation mired in nineteenth-century assumptions.[19]

Domestic reconstruction, along progressive lines of course, thus occupied the editors almost from the start of American involvement in the war. William Hard, the author of one preparedness exposé, lamented the imminent return of laissez-faire, noting that "the summer of 1917 saw great piles of undistributed anthracite coal. The summer of 1918 saw no such piles." Concluding that "it is a triumph of organized units over unorganized individuals," he asked pointed questions: "Shall these units be utterly smashed into chaos when peace comes? Why?" One editorial went even further, insisting that the method of the war was its raison d'être: "In the last analysis a strong, scientific organization of the sources of material and access to them is the means to the achievement of the only purposes by which this war can be justified." As of 1918, the postwar blueprint retained all the key words of Croly's and Lippmann's prewar manifestos. The *New Republic* writers looked to the war to assist those who hoped "to meet the threatened class conflict by placing scientific research at the disposal of a conscious social purpose."[20]

....

Primarily in the pages of the *New Republic*, John Dewey, already the foremost advocate of social intelligence, presented his arguments for the socially formative aspects of war. "Used for the ends of a democratic society," he contended, "the social mobilization of science is likely in the end to effect such changes in the practice of government—and finally in its theory—as to initiate a new type of democracy." While he skillfully maintained a distance from brute mechanism and facile cause-and-effect reasoning, Dewey was at least partially guilty of the abstraction of war into just another intellectual challenge, the move that so disillusioned Randolph Bourne. Dewey strove to live up to the scientific ideal and never completely split fact from value, but moral naivete can be detected in this period.[21]

In this position he was not at all alone, for confidence in the social uses of engineering remained high in 1917, when Dewey asked of war the instrumental question: "The more one loves peace . . . the more one is bound to ask himself how the machinery, the specific, concrete arrangements, *exactly comparable to physical engineering devices*, for maintaining peace, are to be brought about." The historian James Kloppenberg has argued that the "architects of social control should not be confused with these theorists of social democracy [Croly, Lippmann, and Dewey]." But the political implications of engineering modes of reason, no matter how democratically or generously applied, are inescapably hierarchical, in part because of the mechanistic referents of this language and in part because of the hubris that held there could be only one correct logic. That a social thinker as formidable as Dewey so energetically idealized technical success as a criterion for social change testifies to the unique power engineering held for American intellectuals in the World War I era. Although Dewey did not fall into the simplistic positivism characteristic of some Taylorites, his philosophic subtleties did not significantly rock the boat in which the *New Republic* progressives sailed confidently into the future.[22]

The magazine's faith in the wartime administrator originated in part with the editors' confidence in Herbert Hoover. From the outset of Hoover's political career, the *New Republic* steadfastly supported and defended the future president from charges of autocracy. The very model of the engineer-administrator, Hoover returned the compliments, calling the *New Republic* the "best-balanced organ of liberal opinion in America" in 1917. Substantial impetus for Hoover's brief 1920 presidential campaign later originated within the *New Republic*'s editorial board. The magazine also figured in a politically formative episode for Hoover. Edward Eyre Hunt, a Harvard Socialist with Lippmann and John Reed, worked his way through the war as a journalist and wrote a profile of Hoover for Croly. Meeting Hoover so impressed Hunt that he went to work for the former engineer as an aide. Hunt, a management reform advocate, also edited *Scientific Management after Taylor*, a collection of Taylorites' essays to which Hoover wrote the introduction.[23]

While the *New Republic*'s subscription totals climbed rapidly during the war years, challenges to the editors' technocratic outlook appeared even within its pages. Mechanistic language described many political reforms in technological terms; institutional equipment, administrative machinery, and the more generalized uses of *efficient* were typical of the magazine of the 1910s. When one reader protested the underpinnings of this linguistic and programmatic mechanism, however, the magazine's responses further revealed its orientation.

The exchange began with a letter from C. L. Vestal of Chicago in the February 9, 1918, issue. Vestal wrote ostensibly in response to one of Hard's critiques of productive inefficiency but soon turned to wider issues. By criticizing the magazine's fascination with "supermen in government," Vestal explicated the essential myth at the heart of technocratic progressivism. "Your seeming desire and readiness to reconstruct our people into a great machine, each individual becoming a cog, to revolve at the touch of a higher hand, seems repulsive to me," he wrote. Probing a tender spot among preparedness advocates, he continued: "The Germans might have had democracy—they preferred efficiency. . . . To organize the nation as one vast machine . . . is to overthrow in that nation the very freedom for which we hope this war is being waged." Vestal gathered that many businessmen "have, more or less vaguely, perhaps, the glowing, golden vision of an America organized to the last minute detail, as a great commercial machine to dominate world trade after the war." He also tied war economics to the larger issue of imperialism, holding that engineering rationality upheld, and did not replace, economic self-interest.[24]

The editors' response admitted that it would be "impossible to deny some measure of reality to Mr. Vestal's treatment of the existing controversy." But, they continued, "our correspondent makes a grave mistake in confusing this demand for the organization of production with the demand for a ruthless and inhuman mechanical efficiency." Proclaiming that "this matter of organizing production . . . is in reality the ultimate problem of modern society no less in peace than in war," the editorialist looked forward to a day when a wartime munitions department "might well develop into a department capable of producing and distributing the munitions needed for the normal peaceful and fruitful life of a whole society." Blind to the inertia of great bureaucracies and deaf to Randolph Bourne's contention that "war is the health of the State," the *New Republic* thus held fast to a conception of politics as the administration of national productivity.[25]

Vestal wrote again, defining his terms and denying the editors' often caricatured portrayal of his argument. Noting that the response did not even mention the issue of imperialism, also discussed in an article by H. G. Wells in the February 9 issue, Vestal claimed that he would be "highly gratified by one of your illuminating editorials" on the topic. He then returned to the nub of the matter, asking, "Is not the great majority of our leaders of thought and finance looking toward mechanical efficiency as the shining goal to be attained by the nation's newly awakened power?" Some journals, including the *New Republic*, advocated what Vestal called "a real democratic efficiency," but he doubted that it was "the kind which constitutes part of the daily mental pabulum of the masses." He concluded declaratively that "the anti-democratic forces in all countries are the efficient forces." The troublesome issues of imperialism and the antidemocratic nature of expert administration never drew a formal response from the editors of the magazine, but oblique references to Vestal's letter appeared in its pages throughout the remainder of 1918.[26]

Even amidst opposition, the *New Republic* thus expressed many fashionable terms and theories deriving their power from applied science. Preparedness and reconstruction, efficiency, expertise, and "social laboratories"—these were the bywords of the moment. Hopes for empirical knowledge of and rational control over a political universe informed and motivated this confident group of intellectuals and journalists. Despite changes already under way that would diminish the magazine's reputation in the 1920s, the first five years of publication were heady times—of influence, of editorial quality, and of popularity—for such an enterprise. Nevertheless, dissent existed within the magazine's circle of contributors. Randolph Bourne, who began the war in close agreement with Dewey and Croly, soon rebelled against the uses to which education and science—two *New Republic* pillars—were being turned. In the October 1917 issue of *The Seven Arts*, a pointed and significant critique of what the author called "war-technique" appeared, representing a thoughtful and thoroughgoing criticism of rational reform's excesses. Bourne's perspective has served to inspire later generations of the American left, but it should be noted that his dissent reversed a much less radical stand he had recently taken in favor of industry-based government.

In the *Dial* of February 22, 1917, Bourne favorably reviewed Charles Steinmetz's *America and the New Epoch*. He asserted that the book presented a program that was "magnificent and far-flung in its implications"; its author "has that rare and suggestive vision of the socialist who is at the same time a great inventive engineer and an active officer in one of our most advanced and successful industrial corporations." Bourne clearly applied the pragmatic test to Steinmetz in order to portray a man who designed powerful electrical devices as an appropriate political theorist for a technologically competent state. The reviewer looked forward to "a national organism, stable, flexible, effective [that] will be achieved where efficiency and democracy will at last strike their proper balance." In such a "Utopian" world, "efficiency would become a spontaneously lived technique." Although his early death prevented Bourne from completing his theoretical critique of the state as an enemy of human values, here he celebrated a national politics premised on giant industrial capitalism. "Our slow democracy," he asserted, "must fail. The new epoch forces this corporation socialism upon us." While Bourne's justly deserved fame rests on his critiques of mechanism, his defenses of it should not be neglected.[27]

Although Bourne aimed his better-known "Twilight of Idols" at John Dewey, the technocratic progressivism dissected therein was closer in tone and content to Lippmann's ideal. Perhaps too eager to play the role of David, Bourne took on the most formidable intellectual Goliath he could find, choosing Dewey for a number of reasons, some of them personal. Bourne apparently changed few minds, for the ideal of a rationally engineered politics, of science-based consensus, and of greater democracy through increased centralization of authority was inflated to even more grotesque dimensions in the war years.[28]

Bourne had decried this tendency already during the debate over preparedness, when he wrote in a review of C. E. Knoeppel's *Industrial Preparedness* that the modern way of "sublimating the raw crudity of blood and hate and destruction into something not only endurable but sublime" was through the pursuit of social engineering: "Pomp and panoply are gone, but in their place we have the glamor of science and efficiency." Discerning the heroic dimension of war managerialism even before it made headlines, he continued: "Blood and hate and destruction are as cunningly veiled as ever in a sensing of war as scientific management, as a high social technique, as a kind of industrial magic." Like Vestal, Bourne understood the connection of technique to imperialism. An efficient war machine would not necessarily "found a brilliant and wasteless civilization," but it would enable a nation "to play fast and loose in the old international competition for world-trade."[29]

In a sequence of four long paragraphs in "Twilight," Bourne attacked the bases and results of social imitation of engineering, calling it instead "instrumentalism," in keeping with his grudge against Dewey. Attacking a "younger intelligentsia, trained up in the pragmatic dispensation, immensely ready for the executive ordering of events," Bourne deemed them acolytes, "upon whom Dewey, as veteran philosopher, might well bestow a papal blessing." But was Dewey—was philosophy—so responsible for the surge of faith in technique? Bourne himself claimed that these men and women were "the product of the swing in the colleges from a training that emphasized classical studies to one that emphasized political and economic values." Professional tendencies within social science, business, and engineering clearly resulted from more than just Deweyan pragmatism.[30]

Bourne saw that these younger intellectuals were fascinated by the war technique; the "interpretive or political side"—dismemberment and dislocation—held far less appeal. The congruence between training and opportunity was propitious: "It is as if the war and they had been waiting for each other. One wonders what scope they would have had for their intelligence without it." Once again single-mindedly placing responsibility with the reigning pragmatist, Bourne saw that "they are vague as to what kind of a society they want, or what kind America needs, but they are equipped with all the administrative attitudes and talents necessary to attain it." While Dewey called "for a more attentive formulation of war-purposes and ideas," he did so "largely to deaf ears." How, then, was he to be blamed, especially when others including Lippmann remained much less sensitive to the question of social ends?[31]

The answer likely lies in Bourne's previous devotion; he was chastising himself but blaming Dewey for the enthusiasm of a too-uncritical youth. "To those of us who have taken Dewey's philosophy almost as our American religion," he confessed, "it never occurred that values could be subordinated to technique." Admitting that Dewey "always meant his philosophy, when taken as a philosophy of life, to start with values," Bourne criticized pragmatists for being "content with getting somewhere without asking too closely whether it was the desirable place to get." People like Dewey who tried to keep value questions alive had been "too bloodless and too near-sighted." Though he mistakenly attacked pragmatism alone, Bourne saw clearly: "The defect of any philosophy of 'adaptation' or 'adjustment,' even when it means adjustment to changing, living experience, is that there is no provision for thought or experience getting beyond itself. . . . You never transcend anything."[32]

In contrast, Bourne proposed that aspiration had to outrun mere technique; the current political realism had "everything good and wise except the obstreperous vision that would drive and draw all men into it." While pragmatic philosophy certainly held sway, Bourne oversimplified yet again. He had the result right, but his explanation suffered from monocausation: "The working-out of this American philosophy in our intellectual life then has meant an exaggerated emphasis on the mechanics of life at the expense of the quality of living. We suffer from a real shortage of spiritual values." Bourne, in short, struck to the core of the void left as rational reformers unreflectively applied engineering's methods and principles to the state. Means outran ends, technique had no purpose except its own existence, and social aspiration became irrelevant for its being irrational. Revising his understanding of pragmatism, Bourne invoked William James to prophesy that "it is the creative desire more than the creative intelligence that we shall need if we are ever to fly."[33]

Bourne anticipated later critiques of the values embedded in casual emulation of engineering methods. These methods contributed noticeably to political ideologies that by 1915 were starting to become liberal in the modern sense. The leading journal in this renewal, the always self-consciously forward-looking *New Republic*, over the years published the work of many reformers who took engineering seriously as a political resource: Alvin Johnson, George Soule, Morris Cooke, Walter Lippmann, and, less uncritically, Herbert Croly and John Dewey. Nearly alone among his former confederates,

Bourne challenged the fundamental assumptions of rational politics, but even he felt the magnetic attraction to empirical standards of control and performance. In the Great War, many others would evaluate government with similarly empirical yardsticks.

SOCIAL SCIENCE AND ITS APPLICATIONS

While the *New Republic* and its associated circle of intellectuals organized the most visible and influential discussions about a rationalized state, other institutions that became important in the 1920s—policy research bodies, certain philanthropic foundations, and new ventures in social analysis— began to develop in this period. Before World War I, academics connected what might be called the practical and theoretical aspects of the social sciences more tightly than they would afterward. Sociology and social work, political science and public administration, and to a lesser extent economics and business schools shared central definitions, personnel, and missions. Although academic social scientists worked diligently to distance themselves from the street-level application of their findings, in the war years some reformers and investigators stressed "scientific" social inquiry, application, and legitimation of existing social arrangements.

. . . .

The movement to rationalize public administration led to the creation of three major institutions in the Progressive period. The first of these, the BMR, was established in New York in 1907. Funded in part by John D. Rockefeller and Andrew Carnegie, the BMR sought to bring efficiency to urban administration by informing citizens and training social administrators. Before he left for Washington to help reorganize the federal government, Frederick A. Cleveland had led the bureau, enthusiastically endorsing overtly Taylorite methods for civic progress. Such methods were not surprising as Cleveland, an accountant, knew Taylor personally. The BMR continued to attract solid talent during the 1910s; Charles Beard headed the institution after he quit Columbia University in the celebrated academic freedom case.[34]

Administrative reformers had also begun to influence federal politics through two other bodies. William Howard Taft formed a president's commission on economy and efficiency in 1912 in an attempt to capture Progressive voters. It added a political facet to the efficiency enthusiasm, one superficially removed from engineering while sharing its premises, language, and wide attention. Cleveland, who headed the commission, sought centralized executive responsibility and greater reliance on nonpartisan experts, measures similar to those advocated by Morris Cooke. Once the Taft commission was disbanded, efforts to maintain the momentum of administrative reform resulted in the formation of what became known in 1916 as the Institute for Governmental Research, seen as the federal equivalent of the BMR. The IGR flourished in the war years under the direction of W. F. Willoughby. It boasted an impressive board of directors including, among others, Felix Frankfurter, Arthur Twining Hadley of Yale, and the St. Louis millionaire Robert Brookings, who funded the institute after the war and for whom it is still named.[35]

Public exposure to the administrative reform agenda derived not only from the usual sources. The publication of the utopian novel *Philip Dru, Administrator* in 1912 exposed readers to a fantastic new ordered society. Written anonymously by "Colonel" Edward House, *Dru* was named for a most unlikely hero with a "splendid, homely face" who would not wed his adulthood love until he was through administering. Several familiar themes reappear. Tradition, for example, can no longer hold sway: "Gloria," says Dru, "we are entering a new era. The past is no longer a guide to the future." Instead, scientific muckraking brings attention to Dru as he "pointed out that our civilization was fundamentally wrong inasmuch as among other things, it restricted efficiency." After a civil war fought against the "interests," Dru ousts the corrupt power holders, including bought legislators, and proclaims himself dictator; he becomes the Administrator of the Republic. Eventually logic wins out, and civil liberties are restored as Dru convinces the public of administrative realities. Reforms of the judiciary, burial practices, immigration, taxation, and other perennial thorns in the paw of the republic allow Dru to step down, and he and Gloria—a social worker, revealingly enough— sail westward into a San Francisco Bay sunset.[36]

Administrative reformers sought to eliminate partisanship in appointments and, more comprehensively, to dislodge "politics" from the practice of government. Their adulation of nonpartisan experts posed a striking contrast to the well-documented sleaziness of machine politics. These public administration theorists shared a logic and methods with the scientific management movement. Both offered a scientific way out of conflict, one mediating labor and capital, the other superseding a corrupt party system seemingly beyond repair. Both dispensed with goals and made the method of management an end in itself. Raymond Moley, in his Ph.D. dissertation written under Beard at Columbia, noted that scientific management was developing in business during the same years that the penchant for governmental efficiency was evolving along "somewhat different lines."[37]

Each wing of the movement reinforced the other to make for a considerable degree of credibility that derived, in part, from the administrators' self-conscious association with engineering. Writing in a symposium on reconstruction, Cleveland explicated the engineering ideal and illustrated the fundamental conception of technocratic politics: "An organization is a machine made up of human parts," so that "the test of efficiency of both the engine and the engineer is found in the horse-power developed and applied to the accomplishment of group results." As a concession to democratic principles, Cleveland added that "those who are interested in results" should be able "to change engineers whenever they desire," without including any nonmetaphorical specifics.[38]

The antidemocratic aspect of the engineering-based model of government appeared quite openly in this discourse. One writer looked forward to a day when democracy could serve science rather than citizens. "If . . . we advance patiently to the acceptance of the experimental attitude and the method of social diagnosis as our basis of action," he claimed in the *Scientific Monthly*, "democracy may presently be safe for scientific standards." Another contributor to the same journal similarly decried the dangers of too much democracy: "Efficiency demands that not only competent but the *most* competent men available fill all positions of importance." With the broad range of problems faced by a modern industrial nation, however, the search for skilled administrators soon encountered an insuperable obstacle: insufficient expertise. Politics, that necessary evil, remained both necessary and evil despite technocratic efforts.[39]

Within social work, the locus of the earliest attempts at social engineering, similar efforts to those in political administration had began. A huge systematic investigation into existing conditions—the Pittsburgh survey—stood as a model for modern reform. Paul U. Kellogg, the director of the project, was later credited as saying that "we are blueprinting the whole community. We are taking the method of the engineer and we are going to see what is here and what we can make of it." A decade later, wartime social workers called their tasks exercises in "scientific social control," when in fact moralistic pursuits still occupied them. Instead of scientifically planning and systematically attacking new social problems, social workers confronted the traditional evils of prostitution, drunkenness, and sexual impurity. Holding that conflict resulted only from misunderstanding, they proceeded on two unrealistic propositions: that any problem could be settled peacefully, and that they could use science to do so.[40]

. . . .

During the 1910s, both public administration and social work began to be supported by private wealth of a new sort. The great foundations, often bearing the names of robber barons, began to distribute money "wholesale," in the words of one philanthropist, instead of on the neighborhood level of traditional relief. There were unmistakable signs of a major transition, wrote Robert Bruère, the director of the Taylor Society in 1918 and the brother of Henry Bruère of the BMR. He saw "a change from alms and welfare work to a socialized business efficiency." For social work, the Russell Sage Foundation played the most active role; it underwrote the bulk of the Pittsburgh survey. In political and economic research, Rockefeller money came to be more important, especially in the 1920s.[41]

Two episodes involving the Rockefeller Foundation illustrate the many complications involved in aspirations toward scientific social research and application. In 1912 John D. Rockefeller, Jr., had met with some of the most powerful Americans of the day—Theodore Vail of AT&T, J. P. Morgan, former senator Nelson Aldrich (Rockefeller's father-in-law), and others—to discuss the formation of a bureau to dispense what they considered sound economic judgments. In Rockefeller's words, they saw an "urgent need for some intelligent, well-conceived, broad, non-political, effort to educate public opinion." The supporters soon differed, however, as to the balance between scientific objectivity, to be achieved by hiring leading economists, and public dissemination of orthodox findings. They saw no contradiction, however, between calling their enterprise nonpolitical and asking for $1 million a year in hidden contributions.[42]

This effort came to naught several times early in the history of the Rockefeller Foundation, but the institution did influence public policy by its involvement with the IGR. During a later and delicate discussion of how a private philanthropy, one vulnerable to public opinion, should affect public policy, a trustee recalled a precedent. "One of the earliest activities of the Rockefeller Foundation," announced Jerome Greene, occurred "in 1915 or 1916 [when] we gave the initial impetus to the Institute for Governmental Research." The results could not have been better: "We were in a position— that is, the Institute of Government Research was in a position to set up the budget legislation of the United States." It did so by writing the bills in Congress, drafting the first budget of the United States, and supervising "the carrying out of the first two or three budgets." In sum, Greene said, "the prime mover in that whole thing was the Rockefeller Foundation, but we worked through the Institute of Government Research." The Rockefeller philanthropies, always in the background, continued to support the collection and application of social scientific data throughout the interwar years.[43]

Within the academic social sciences, as in social work and municipal reform, hope in science ran high. Prominent leaders in economics, political science, and sociology united democracy and science in aspirations toward peaceful social transformation and progress, both before and during the war. All three disciplines moved away from this type of reformism in the 1920s, but during the war years positivism and quantification mixed easily with social hope and technocratic promise.

Among economists, Wesley Clair Mitchell (1874–1948) of Columbia built support for rational reform through his development of institutional economics and extra-academic institutions for research. His book of 1913 on business cycles broke new and acclaimed ground, and work with war statistics led to the formation of the National Bureau of Economic Research in the early 1920s. Mitchell's scholarship had one goal, according to his wife, Lucy Sprague Mitchell. "All his life he campaigned for a method of studying social ills which would yield tested knowledge of actual conditions," she wrote in her memoir of their life together. "Only such knowledge of interacting social conditions, he believed, would equip people to think clearly and to plan social reorganization or 'reform' intelligently." In one address, Wesley Mitchell drove home the point, metaphorically calling society a machine: "Reform by agitation or class struggle is a jerky way of moving forward, uncomfortable and wasteful of energy. Are we not intelligent enough to devise a steadier and a more certain method of progress?" Mitchell had embraced the efficiency discourse—literally—and also endorsed eugenic solutions. Rational reform extended to all areas of life.[44]

Professors of sociology stressed statistical inquiries focused on both efficiency and social control, especially under the influence of Franklin H. Giddings and Luther Lee Bernard. F. Stuart Chapin, a student of Giddings's, asserted that the discoveries of academics and social workers would gradually become part of "that body of tested scientific principle . . . needed to solve the pressing problems of our democratic social order." According to Charles Ellwood, sociology needed to develop "adequate machinery" to instill "rational likemindedness and a rational will in the group as a whole." Education provided the means, Ellwood believed, to make democracy as efficient as autocracy. Once "the masses have been taught to play the social game and to play it well, they will be *more* efficient"—just like a good football team in which "every member waits upon direction from above before he plays his part." The clearest and earliest call for social engineering came from Bernard, who in 1911 claimed that social facts "should become as obligatory as the laws of astronomy or physics" and that "every social organization must be coercive to the extent necessary for efficiency or it must break down."[45]

Among political scientists, similar themes appeared, especially among the public administration scholars already examined. Along slightly different lines, Jesse Macy, president of the APSA, delivered a 1917 address appealing to the growing prestige of pure, rather than applied, science. Idealized engineering thus recedes slightly, but familiar premises—the rationality of cause and effect, truth found in method, and contention as an evil to be overcome—reappear. Macy reintroduced virtue, recently the wallflower of political theory, in updated dress: "The modern scientific spirit is simply the Christian spirit realized in a limited field of endeavor." The leap from the laboratory to city hall was short; these were "two fields of science— one dealing with oxygen, hydrogen, and gravity; the other with cities, states, and numerous other political and social institutions. Both furnish occasion for the exercise of the same spirit and method." That spirit would reshape politics, holding out the possibility that differences could be rationally settled. "Scientific Debate thus became the model and method," Macy claimed, "for dealing with all questions on which men differ in opinion." This understanding of science provided the model of a world where things were what they were: "There must be agreement in definitions, else there can be no proper debate." Even though he appeared to ignore the politics of definition, Macy nevertheless attempted to retain a strong democratic element in his application of science.[46]

. . . .

The implications of this growing reliance upon technological models for social reform multiplied in the 1920s. Innovators who developed their ideas in the war years saw in engineering a powerful example of scientific manipulation of existing structures and materials. Managerial practices, many originating within mechanical engineering, further reinforced the need for decisive administrative solutions to the magnifying and multiplying problems of a technological civilization. Between 1912 and 1920, organs such as the

New Republic provided a forum for the intellectual working-out of engineering and managerialism's promise for a changing world, and new institutions further organized "social intelligence." While World War I implemented the new methods, its carnage and complexity challenged many reformers' rational confidence.[47]

IMPLEMENTATION AND REDEFINITION

1 9 1 8 - 1 9 3 4

One's conviction that the evidence one goes by is of

the real objective brand is only one more subjective

opinion added to the lot.

WILLIAM JAMES

4

WAR AND RECONSTRUCTION

The war to end all wars became a war to concretize progressive administrative reforms as well, and nobody emerged from the conflict untouched. Some radicals became markedly antidemocratic, while many moderates disowned the cause of reform. Certain forces of reaction, stymied by the breadth of progressive aspiration for so long, came to the fore. The analogy between social science and natural science, meanwhile, remained sturdy and adaptable for academics, engineers, and journalists needing inspiration and credibility during the war. Yet the conflict challenged illusions of expertise, rationality, and other forms of social control. Not only was the machine simply too powerful, diffuse, and illogical for any group to steer it decisively, but war's bloody devastation made cool-headed retreat from ideology and conflict impossible to attempt. These discoveries shook some of the confidence from advocates of technocratic politics, so rational reform assumed new forms in the 1920s.[1]

WORLD WAR MOBILIZATION

Middle-class technicians and academics attempted to use preparedness and, later, mobilization as avenues to achieve domestic and professional objectives. Engineers, psychologists, and other medical and scientific leaders understood the issue of war readiness as a problem to be solved, and each profession thought itself uniquely suited to lead the way. All these groups proposed industrial standardization, scientific management, and organized research. For all the conviction these momentary technocrats brought to the war, theirs was not the model of mobilization ultimately implemented at the highest levels. Instead, Woodrow Wilson navigated a middle course, fending off supporters of laissez-faire on one side, engineers and managers on the other.[2]

While personally opposed to the technocratic model, in crisis Wilson did not hesitate to proclaim its advancement of administrative competence. When he named the Advisory Committee to the Council of National Defense, Wilson announced that the members, "appointed without regard to party, mark the entrance of the non-partisan engineer and professional man into American governmental affairs on a wider scale than ever before." Industry simply could not and would not respond to rational control, however, and the technicians, for all of their expertise, lacked aggregate economic data, inventory records, and production schedules to undertake effective coordination. Invested with little authority as compared with the antiquated but entrenched army procurement establishment, the ACCND aroused congressional pressure with its less than spectacular success. In late July 1917 the War Industries Board came into being, but it too had to be reorganized in the following year, when Bernard Baruch took control over a vast octopus with little formal power. Nevertheless, the centralization of the control in the executive branch of the federal government meant that *The Promise of American Life* had, in partial measure, come true.[3]

Holding the most powerful presidential mandate in American history to date, Wilson resisted the technocratic imperative to a substantial degree. Needing to transform a citizen army—and economy—into substantially more professional and proficient variants, he relied primarily on calls to patriotism and voluntary sacrifice. More cautious than either Abraham Lincoln or Franklin Roosevelt, Wilson chose citizen war councillors who reflected his ideology. None of the members of the WIB could be called a thoroughgoing technocrat, and the ideal they pursued was closer to that of a gentleman mechanic than a rational manager. Within the multilayered administrative structure of the WIB, and inside industry, however, the gentry model failed, and "compromises" between public and private interest did occur.[4]

To reconcile technocratic managerialism and a laissez-faire tradition and to avoid both autocracy and industrial anarchy, the WIB made highly utilitarian decisions. Relying on cajolery, exhortation, intimidation, and negotiation, the board employed more of the salesman's methods than the technician's. It mingled administrative coercion—usually threats of negative press —and old-boy connections in its pursuit of a minimally coherent industrial policy. The admixture thus could be called neither private enterprise nor public planning; instead, paradoxes, compromises, and cobbled together arrangements characterized the WIB. After the engineers and industrialists on the ACCND failed to orchestrate an entire economy, Baruch, the intuitive Wall Street investor, proved the perfect man for such an ambiguous job. But even his attempts to "cultivate public taste for rational types of commodities," while coercive enough to unsettle many Americans, could not accomplish such a rarefied goal. The speculative capitalist who finally made the WIB work came from outside engineering, administration, and managerialism, the arenas of such unproven self-confidence.[5]

The World War I mobilization failed to integrate the economy, predict and organize consumer preference, and unclog railroads frozen by ineffective management and overwhelming demand. Still, an undertaking of such enormous proportions raised technocratic hopes within many observers and participants. Among those intellectuals and technicians involved in war activity, Edwin Gay, Walter Lippmann, and Herbert Hoover represent important junctures of ideology, opportunity, and capability in the transformation of prewar reform into a comprehensive understanding of social control. None of these men enjoyed unqualified success, but each implemented some sort of social engineering and contributed to a mythological understanding of war as a sterile abstraction.

Edwin Gay, the former dean of the Harvard Business School, became the most powerful of the many Taylorites who flocked to the duties of war management. His comrades Morris Cooke, Henry Gantt, and Henry Dennison worked to get American shipping expanded and organized. Harlow Person, Carl Barth, and H. K. Hathaway served in ordnance positions. Ida Tarbell was a member of the Women's Defense Work Committee on the Council on National Defense, and Frank Gilbreth was a major in the Engineers Corps. Henry Kendall sat on a committee of surgical dressing manufacturers. Gay began his war service on the Commercial Economy Board, another offshoot of the Council on National Defense, working under Arch Shaw, the publisher of *System: The Magazine of Business* and an ally of Hoover. The board, which also included Gay's Taylor colleague Henry Dennison and three other members, worked to "investigate and advise in regard to the effective and economical distribution of commodities among the civilian population." Its members sought to convince manufacturers and retailers, in the words of Gay's biographer, to "abandon the uneconomical." Gay's specialty was clothing: Could the number of shoe styles be reduced? Could coats be made with less wool? He found work on the board frustrating insofar as he, like Baruch, was forced to rely on persuasion with little tangible authority.[6]

In February 1918 Gay was named to head a division of planning and statistics within the Shipping Board. The army needed to transport 600,000 men to the front in the first half of 1918. To free up enough ships, imports and exports had to be drastically restricted, and the quantity of information the statisticians attempted to coordinate remains mind boggling. How many sausage casings did American packers need? How much cargo space did that quantity require? Such seemingly trivial but ultimately momentous decisions strained the administrators, and in confronting the heaviest of the Washington heavyweights—Hoover and Baruch among them—Gay acquired what he called a "bellicose" reputation as "a good deal of a bull in a china shop" as he attempted to make the necessary vessels available.[7]

In addition to having to schedule shipping, Gay had been asked by Baruch in June 1918 to organize the Central Bureau of Planning and Statistics. He was therefore responsible for a "conspectus of all the present war activities of the Government" that Wilson could periodically review. The Central Bureau, a clearinghouse and advisory office, produced no statistics but collated, cross-referenced, and coordinated the efforts of other agencies. After a burst of energy expended in arranging economic data for Wilson at the Paris peace talks—data the president largely ignored—the Central Bureau was on the verge of being disbanded. Gay, however, held out hopes that it could be sustained as a peacetime agency and even threw what he later called a "disgraceful" and "childish" tantrum in Baruch's office. Despite his close weekly reading of the bureau conspectus, Wilson insisted on rapid demobilization, and the Central Bureau's budgetary appropriation lapsed on June 30, 1919. In 1920 Gay helped to found, with Wesley Mitchell, the NBER.[8]

The ideal of expert scholars, adept at interpreting data and offering practical recommendations, found numerous expressions in the war. One such group, the Inquiry, operated at the highest, most secret, and most ideologically charged level of American politics. Edward House led this investigative team whose ostensible mission was to produce data for Wilson at Versailles. Among the members of this elite cadre was Walter Lippmann, who had quit the *New Republic* in June 1917 to work for the War Department. After serving as an assistant to Secretary of War Newton Baker, Lippmann was named to the Inquiry—possibly to placate Wilson's liberal intimates— and drew a salary second only to James T. Shotwell of Columbia, the director. Somewhere along the way, Lippmann apparently unwillingly supplied Wilson with the slogan about making the world safe for democracy. The history of the Inquiry reveals similarities with other wartime agencies. For all their credentials, the experts ended up writing on topics far outside their fields, and the data, while voluminous, often were given no more in-depth treatment than the interpretations found in encyclopedia articles. Language barriers, antiquated sources, and staff shortages made expertise little more than a self-congratulatory euphemism. In the spring of 1918 Lippmann quit to work for military intelligence in Europe.[9]

Throughout this period, Lippmann had his eye on the larger implications of wartime behavior. What he saw forced him to revise his positions, so forcefully argued in *Preface to Politics* and *Drift and Mastery*, on science and citizenship. As of 1915, in *The Stakes of Diplomacy*, Lippmann's hope in Progressivism still remained intact. Asserting that "the people have suffered, worked, paid, and perished for ends they did not understand," he saw reason for optimism on the political horizon. "There is on foot a highly intelligent movement," the book declared, "to reconstruct political machinery so that government becomes visible and simple and responsible." The upshot was the hope that "the technique of government may be far enough advanced to allow wider and wider groups to take part in the affairs of diplomacy." Here again, social intelligence, plus responsive and rational government, equaled modern democracy.[10]

Liberty and the News, published first as essays in the *Atlantic* in 1919, revealed this transition in the making. Less articulately and completely than *Public Opinion* (1922), *Liberty* challenged the old beliefs in the promise of scientific method among a newly informed and empowered citizenry. Lippmann contested the ability of citizens to fashion informed judgments, decrying a "breakdown in the means of public knowledge." "Increasingly," he wrote, "[people] are baffled because the facts are not available, and they are wondering whether government by consent can survive in a time when the manufacture of consent is an unregulated private enterprise." Citizens mocked scientific methods as "men cease to respond to truths, and respond simply to opinions." Since "the really important thing is to try and make opinion increasingly responsible to the facts," the BMR, the IGR, and other institutes promised improvement. They could serve as "political observatories" where experts capable of statistical and other advanced forms of understanding, untainted by mere opinion, could condense the news into a factual form ready to shape action. All in all, in *Liberty and the News* Lippmann began to doubt the "informed citizenry" he so thoroughly demolished three years later in *Public Opinion*.[11]

Lippmann himself played an active role in the manipulation of the public whose capacity he so doubted. In 1917 he had written Hoover with suggestions for a propaganda campaign. Declaring himself "yours devotedly,"

Lippmann asserted that "the motives that are to be worked are patriotism, social pressure in the local communities, the sense of what is respectable and what isn't, and finally, the threat that drastic powers are in reserve." Lippmann had also dined privately with Wilson, and one wonders about the motivations for his actions at this juncture. Having access to power became a habit, and the most oracular prophecies were the self-fulfilling ones.[12]

Of the three figures under consideration here, Herbert Hoover moved from wartime to private life with perhaps the most positive reputation among the public at large. To Hooverize was the essence of efficiency; the term connoted standardization, reduction of waste, and patriotism, all achieved voluntarily. But Lippmann's complaint that "the manufacture of consent is an unregulated private enterprise" failed to address the systematic public maneuvering carried out by war agencies, foremost among them Hoover's Food Administration and William Gibbs McAdoo's Treasury Department. The voluntarism of which Hoover was so proud resulted largely from deliberate, heavy-handed, and often naive application of advertising and motivational devices. While public sentiment spun out of control and lynchings and other mob frenzies blighted the land, Hoover remained—justifiably—proud of his department's achievement in feeding citizens, soldiers, and war victims with staggering effectiveness. At the time, though, such connections between propaganda and performance escaped many observers.[13]

Hoover's food duties began with his conduct of the Belgian relief efforts, after which he sought, and was given, extensive authority to control American food production and consumption. A reputation for forcefulness preceded him; Colonel House told Wilson that Hoover was "the kind of man that has to have complete control in order to do the thing well." From his staff, Hoover commanded great loyalty with a capacity for endurance and agile manipulation of reams of data that was, by all accounts, prodigious. When he appeared before the Senate Agriculture Committee, Hoover argued for the many qualities beans possessed as healthful, protein-rich food. When a senator asked him about the price per bushel, Hoover replied that he didn't know. "I have always bought them by the ton," he said. Conceiving of the nation—and to some extent, the world—in aggregate terms, Hoover sought to mobilize American farmers, housewives, and food processors to participate with minimal formal coercion. In his quest, Hoover made many enemies, being compared by one senator to Caesar "in the bloodiest days of Rome's bloody despotism."[14]

The strikingly colorless moral universe Hoover created reveals the extent to which technique dictated value. In a letter of 1918, he posed the issues involved with his unprecedented authority in bland, problem-solving terms: "As Food Administrator, I have no general reforms, no spiritual movements to undertake, but simply a purely practical end to attain. . . . There are infinite injustices and wrongs in the United States and an infinite amount of social evils. Like every other citizen who loves his people, I would truly like to see these things remedied. But it is a job that I cannot undertake and at the same time successfully fill my niche in prosecuting the war." The letter constitutes one of the most revealing of Hoover documents. Little given to reflection, he favored facts, action, and consensus, preferably with him in the lead.[15]

Hoover fit well into Wilson's conception of a temporary wartime gentry and without question performed his job well, holding in mind explicit plans for the demobilization of what was from the outset understood to be an emergency agency. His political economy in the 1920s would be shaped by the complementary concepts of service and voluntarism, with a definite elitism implied: society's most competent and visionary, not just its mediocre masses, would heed the call to serve. Generalizing from his own experience, he attributed America's many successes to the concept of equal opportunity, assuming that other elites would do three things, among others: embrace the ideal of service, bring definite skills to their public duties, and remember their origins, helping to pave the road of success so more could follow. The wealth-crazed 1920s exposed such notions as open to challenge, for Hoover was more of an exception than even he, a man without undue humility, realized.

Although Hoover, Lippmann, and Gay diverged during the 1920s, like many of the wartime intellectuals they retained the war's lessons about political and public rationality. While doubting the possibility of large-scale centralized control of production and distribution, these reformers believed education, social research, and expert leadership were crucial to the modernization of American democracy. All three men looked to the newspaper as a tool for social education, as did other rational reformers: Gay edited the *New York Post* for a brief while, Hoover played with the idea of buying a newspaper, and Lippmann abandoned full-time magazine journalism after 1920. (Later, the Twentieth Century Fund worked to make newspapers "better aids to democracy," while Beardsley Ruml suggested that the Rockefeller Foundation buy, merge, and operate the *New York Times* and the Scripps-Howard newspaper chain.) The war brought administrators of many sorts together in an intense, purposeful, and unforgettable environment. Afterward, managerially minded businessmen like Henry Dennison and Arch Shaw, so-

cial scientists including Wesley Mitchell and Ruml, and future philanthropists such as Raymond Fosdick and Frederick Keppel maintained contact. The network of war bureaucrats helped develop and motivate organized
social scientific research in the following decades.

AFTERMATH AND RECONSTRUCTION

The Great War entrenched technocratic progressivism in American political
culture, if not in the federal government. Hoover's brilliant administration of
Belgian relief efforts and American food allocation, the recruitment by the
federal government of thousands of technicians, and the discrediting of dissent all fed a curiously apolitical reform spirit. Less publicly engaging than
Theodore Roosevelt standing at Armageddon and fighting for the Lord but
coexisting with such expressions of uplift zeal, a coalition of social scientists, engineers, journalists, and philanthropic foundations doggedly pursued technocratic programs of reform in the 1920s. The sheer size and structural complexity of modern warfare convinced many that moralistic reform
was no longer adequate in peacetime. Social structures must be measured,
these intellectuals believed, and then shaped by the tools of modern management and social science. Postwar liberalism blended a preexisting belief
in social perfectibility with a growing confidence in sophisticated methods of
analysis and coordination. Administrators could solve the problems of the
era, proposing politically safe solutions often by denying that politics, in the
old pejorative sense, even existed. By 1928 a president would be elected
whose supporters included some who claimed that he was not a politician.

Despite the attention paid to figures such as Hoover and Gay, the locus of
reconstructive expertise shifted after the war as the federal government refused to maintain most of its experiments in managed social change. Two
particular developments altered the shape of emerging semipublic intellectual institutions. First, psychology became the social science of choice
among technocratic reformers; for some, the society as machine gave way to
the person as machine. Along with engineering, chemistry, medicine, and
anthropology, psychology was defined as a science by the National Research
Council; sociology, political science, and economics were not. Social engineering remained a central component in social thought, but now the enterprise diverged. The famous intelligence testing program conducted by Lewis
Terman, E. L. Thorndike, and their colleagues led many theorists to think
adjusting the person to the state appeared more likely than successfully con-

trolling anything as vast as a nation-state. According to Terman, "If the Army machine is to work smoothly and efficiently, it is as important to fit the job to the man as to fit the ammunition to the gun." George Patrick, a eugenics advocate, declared that "the world will be made safe for democracy only when the people of the world are made fit to live in a democracy." Behaviorism's mechanistic assumptions, in intellectual vogue during the postwar period, further reinforced those who favored the engineering motif.[16]

Second, because they had been home to social scientists, engineers, and military training centers, American universities emerged from the war transformed. During the war, academics had concluded that their social function had become the advancement of the state's pursuit of military victory. This position ran headlong into the standards of detached social criticism espoused so assiduously by the social scientists, but resulted in added prestige for the war's brain workers. Higher education developed ways to participate in, rather than critically assess, the task of making technological America more efficient and prosperous. After the war, the universities expanded their role in this transaction. Edward T. Devine, an influential social worker, wrote that a world awaited where "all shall have enough to live on, education enough to know how to live, and health enough to enjoy life." Charles Ellwood, he of the democratic football team, foresaw that "the work for rational and scientifically planned social progress lies all ahead. And socialized education is the key to social progress." Within an ideology stressing education in broader terms, universities both trained experts and taught the public to appreciate those experts' contribution.[17]

New institutions like the research university became necessary because, as observer after observer attested, times had changed permanently and decisively, and old modes of conduct and beliefs no longer pertained. This rage for the new characterized political writing, which referred to the New America, the New epoch, the New era, the New nationalism and freedom, and the New machine. (Significantly, within political language, the new was the rational, the scientific, the mechanical. In the arts, the modern was the disjointed, the illogical, the uncertain.) The *New Republic* claimed that "during the war we revolutionized our society." One journalist "saw a new order for America—just ahead." Woodrow Wilson told the nation that "the world will never be again what it has been. . . . We are provincials no longer." Charles Ferguson's fictitious engineer recalled that as he "witnessed the Great Mobilization" of 1916, he "was at the gate of the New Age with those that wedged it open. I helped turn the hinge of universal history."[18]

In line with these declarations of epochal transition, the end of a gruesome, apparently pointless war and the failure to devise a rational forum to maintain world peace renewed convictions that reform along scientific lines had become imperative. The savagery of mechanized and chemical warfare shocked many who had seen only humanitarian uses for advancing technologies, and they came to endorse what Veblen had said long before: political institutions failed to keep pace with human technic. The logical solution was to remake government in the image of engineering, the source of the technology that had outmoded social institutions in the first place. Raymond Fosdick, who as a trustee of the Rockefeller Foundation oversaw an aggressive program of "social science and social technology" in the 1920s, told a graduating class of college seniors that the war years had shown "the abyss upon the edge of which the race is immediately standing." He foresaw "inevitable doom" unless "we can achieve a measure of social control far greater than any which we have hitherto exercised." The disillusionment of the war, rather than leading to despair, spurred "a determination through some means or other to speed up the development of social controls, to bring men ethically and morally abreast of their own machines." Fosdick's own experience in the war bureaucracy convinced him of the inevitable need for scientific styles of reform, and many of his comrades emerged with similar outlooks.[19]

Because the war mobilization, "the biggest social machine of all," had transformed civilization, technocratic reformers reached the inescapable conclusion that a similarly mechanistic paradigm would inspire the government of the new civil order. One political scientist noted that "in the demands which it made upon democratic government the war operated as a supreme test which revealed flaws in the machinery not otherwise noticeable." Robert Bruère contended that "eighteenth-century governmental machinery was not designed" for such a huge task as world war. The existing political system was "like a wooden mill-wheel caught under the falls of Niagara, where dynamos and turbine engines are needed." Advocates of such a politics argued that the forces that had caused unprecedented misery had to be transformed, and not challenged, for humanity to have a chance.[20]

To effect such control, the possibilities for a peacetime industries board appeared manifold; Grosvenor Clarkson, the in-house historian of the WIB, referred to "dreams of an ordered economic world." "Some day," continued Clarkson, "it may occur to some President to apply the organization scheme of the War Industries Board to Government." Status, for a moment, had come to rest on the shoulders of administrators; Baruch, Hoover, and McAdoo, not the military figures, ended up being the heroes of the war. Their disappointment at being brushed aside, at not being given a chance to apply the findings of the war, lingered afterward. Rexford Tugwell, by 1927 an economics professor at Columbia, recalled that "there were none who thought the competitive system sufficient for war." In the end, he argued disappointedly, "we were on the verge of having an international industrial machine when peace broke." The war, "an industrial engineer's Utopia," had conferred insights too soon forgotten. "Perhaps we shall turn back to these pages some time for a reformulation of industrial policy," but not in the Coolidge era. "War is war," he bitterly reasoned, "and peace is only peace." The market model, not a peacetime industries board, returned to ideological primacy.[21]

For all the misty-eyed recollections, confounded designs, and emptied offices, hopes for technocratic war management outstripped its performance. Despite the rhetoric, the federal government lacked trained personnel, adequate information, and sufficient experience for economic regulation on the scale demanded by the war. The WIB, meanwhile, successfully maintained the myth of a coherent system that never existed. Like the Inquiry scholars, wartime technocrats tried to look as though they mastered their surroundings, but they exercised little broad control over the leviathan. Gay's and Hoover's experiences illustrated this inadequacy: cajoling, praising, and threatening all the different sectors of a newly integrated economy would not have worked were it not for the "patriotic" war frenzy in which they operated. Steering that political beast once they inflamed it to action proved more difficult yet. The temporary cross-breeding between laissez-faire and a command economy, in short, showed that the American technocratic dream was an illusion. Like most illusions, it died hard among its most ardent believers.

. . . .

In the vanguard of these erstwhile advocates, Herbert Croly and the *New Republic* writers retained their concern for rational reconstruction during the war years. In 1918 Croly elaborated on the connections between academic expertise and political progress and proposed new institutions to train qualified men and women. In writing that "modern society is undergoing a process of quick and radical transformation, which most of its official leaders are insufficiently prepared to understand and control," he restated yet again the social engineering credo. Acquisition of "the technique of social progress" could best occur not in the modern university but in a research setting where the "intellectual energy of its staff [could center] upon the study and mastery of social processes." He called for schools of social science to train social administrators. This new type of institution would "gradually give to the work of social administration, engineering and research a professional standing similar to that now enjoyed by physicians and lawyers." Once again, professional self-interest and putative public benefit cohabited within the same ideology.[22]

The proposal became reality as the New School for Social Research opened its doors in 1919. Beforehand, planning sessions that took place at the *New Republic* offices had included many standbys of both the magazine and American technocratic progressivism: Thorstein Veblen, Felix Frankfurter, Wesley Mitchell, Alvin Johnson, and Walter Lippmann. John Dewey and Charles Beard came into the fold later. Most of the founders hoped that social scientists, freed from politics, tradition, and religious rationalism by the spirit of scientific inquiry, could use technocratic methods to reconstruct America.

The New School's self-justification soon came under attack, implicitly from Bourne's legacy on the left and vociferously from critics like Nicholas Murray Butler on the right. Butler, the president of Columbia who lost several key professors to the upstart academy, slammed the founders as "a little bunch of disgruntled liberals setting up a tiny fly-by-night radical counterfeit of education." The understanding of technology among some of the faculty, especially Veblen, no doubt fueled the ire of the critics. "We once considered calling the school The Institute of Social Technology," wrote one student to the *New Republic*. "I think that might have been a pretty good name. The Technology would have reassured those who were scared by the Social." Pragmatism, so comfortable alongside mechanism, prevailed: "Consistently we aim at making rapprochements which will really illuminate the social process and at testing knowledge by constant reference to life." Despite the rhetoric, the obstacles to the realization of Croly's New School mission prevailed, and the first incarnation dissolved, amid contentiousness and mismanagement, in 1922. Revamped under Alvin Johnson, it took a new direction away from its original ideas.[23]

The 1920s proved to be a difficult time for Croly and the *New Republic*. A flirtation with technocratic politics—the brief presidential campaign of Herbert Hoover—came to naught as persuasive letters opposing the magazine's endorsement appeared frequently. (One contended that Hoover's only attribute was his "unquestionably efficient record" in the war; but "nine-tenths of the suffering in the world at the present time is due to efficiency, selfish efficiency, devoid of moral purpose or rational direction.") Clearly Croly labored under the double burdens of fatigue and disappointment, and no great synthetic book wove his postwar beliefs into a pattern, orderly if not vivid. What writing he did publish consistently refuted the drive toward centralized administrative rationality, emphasizing instead the mystical and spiritual; he decried his previous efforts at rational reform as a "mistake." In 1925 Croly damned explicitly much of what he had stood for. The hero of the new sociology, "the social engineer," he wrote, "tended to become in practice a revised edition of the traditional law-giver who knew what was possible and good for other people and who proposed to mold them according to his ideas." Reform, for some, meant "placing at the disposal of social engineers a machinery of economic, social or legal coercion." This revised image of the machine represented not liberation and empowerment but "coercion" and oppression; the engineer was rhetorically transformed from savior to slavemaster.[24]

Most of Croly's *New Republic* pieces defected from the realm of technique and rationality he helped inspire before the war. A fragment of an unpublished book published in the magazine overflowed with almost unintelligible technospiritual warmth; once "religious people act immediately and courageously on what they have learned" from scientific methods, human nature could "unfold itself with unprecedented momentum." Elsewhere he argued that "liberalism is an affair of the spirit and resides, if anywhere, in the human soul." By the time he formulated "The *New Republic* Idea" in a special pamphlet designed to increase subscriptions, Croly abandoned social engineering for the "soft" rationalism so repugnant to technocratic progressives. Democratic discussion "needs to be transfigured by a common conviction of the latent regeneracy and brotherhood of mankind," he wrote. "It means the worship of a God symbolized not as Power, but as Understanding and Love." The shaman had replaced the technician, sentimentalism the scientific.[25]

. . . .

Other reformers retained their confidence in political technique, implicitly or explicitly invoking the engineer. One mechanical engineer wrote in 1919 that "we fear no political backfire. We have no fences to mend. We can stand in the open and say everlasting truths, and the time will come when some men believe them." In a similar tribute, Samuel Gompers called engineers "the scouts of civilization" in 1920. Secretary of Interior Franklin K. Lane, an archetypal progressive who had warned against undue reliance on experts during the war years, argued a different and far-reaching line in 1920. After modern life grows in complexity and government broadens its activities, he asserted, expertise becomes necessary: "We are compelled to employ the engineer. And by the engineer, I mean the man who can apply imagination to the facts—the planner, the one who sees his way through." Lane expanded the definition far beyond professional realities into the realm of the metaphoric. "The men who drafted the constitution of the United States were engineers," he claimed. "Our great pieces of legislation, such as the Federal Reserve Act, are matters of engineering just as definitely as an irrigation project. . . . The Congress is the engineer of the nation."[26]

In March 1919 William Henry Smyth codified the war machine's political assumptions into a plan for "National Industrial Management," which he called Technocracy. Founded upon the engineering model, Technocracy implemented science in the drive for control, which, following Lippmann, Smyth called the Mastery Instinct. Once again, the war was seen as a decisive turning point; "indeed . . . our great but chaotic nation—in self-preservation—*ceased to be a Democracy*!" After a "most remarkable transformation," the United States did not become an autocracy, a plutocracy, or even a theocracy. Instead, he claimed, "during this thrillingly interesting time, the United States developed into a form of 'Government' for which there is no precedent in human experience." In a new and higher phase of social development, "we rationally organized our *National Industrial Management*. We became, for the time being, a real *Industrial Nation*." Smyth glorified war technique in his account of political transformation: "*This we did by organizing and coordinating the Scientific Knowledge, the Technical Talent, and the Practical Skill of the entire Community; focussed them in the National Government, and applied this Unified National Force to the accomplishment of a Unified National Purpose.*" Smyth wanted to "organize our scientists, our technologists, our exceptionally skilled" to unify them for the public good. These experts, "a Technical Army devoted to Peace and Construction," would "facilitate the full and socially useful outflow of the three vigorous forms of life energy—*Strength, Skill, Cunning*." Smyth thus spelled out a name for and a description of what many had hoped for from the war. Unlike *efficiology*, suggested at about the same time, *technocracy* has endured in the lexicon.[27]

The most noteworthy attempt to politicize engineers in this period proves more difficult to decipher, coming as it did from a mind of great acuity, reclusiveness, and originality. When Thorstein Veblen attempted to mobilize engineers to overturn pecuniary capitalism, he moved further into the intellectual and programmatic mainstream than ever before. His motivation for doing so remains unclear, perhaps ambiguous even in its germination. While at the New School, Veblen taught a course on industrial management, including works by Henry Gantt and L. P. Alford (later Gantt's biographer) on the syllabus. After he read some work by Morris Cooke and Gantt—who derived his ideas largely from *The Theory of the Leisure Class*—Veblen mistakenly considered the engineers to be ripe for political action, appraising them to be "in a position to make the next move." Cooke had asked his colleagues "How about it?" in one of his pamphlets; Veblen ominously portrayed the engineers drawing together and asking "What about it?" of a program more radical than any of Cooke's internal reforms. Veblen's early pieces reprinted in *The Place of Science in Modern Civilization* (1919) include some of his most astute and complex work. With those essays so close at hand, why the "technological hallucination," as Samuel Haber calls it, of *The Engineers and the Price System*? Daniel Bell plausibly suggests that the essay may have been a joke. Phrases like "massed and rough-handed legions of the industrial rank and file, ill at ease and looking for new things" sound propagandistic and yet tongue in cheek at the same time. Surely Veblen knew what he was doing.[28]

Or did he? *The Engineers and the Price System*, so removed from Veblen's other work, may best be understood as the incongruity that got away. As David Riesman discerned, there are times when "Veblen himself, caught in the paradoxes of his irony, appears to be unsure what he is mocking and what he is glorifying." In the place of the earlier playful yet serious subtlety appears a brittle bitterness; the toll of a life lived on the outside looking in may have been the supple confidence of a great but isolated mind. Lewis Mumford, an admitted disciple, saw this at the time. In the book, he wrote, "something has been gained and something lost: what has been gained is an insight into a certain weakness in Mr. Veblen's philosophy; what has been lost is that delightful turn of humour which so thoroughly concealed the weakness." When Veblen became prone to utterances like "the common man has won the war but lost his livelihood," he could be chided but excused for momentarily losing his touch. When he was blatantly irresponsible, charity comes harder. The cheaply foreboding refrain "at least just yet" taints the book; so does the statement that "quite an unusually large number of machine guns have been sold to industrial business concerns of the larger sort, here and there, at least so they say." Many passages were intended, it seems, to cost Veblen respect.[29]

One factor remains damning, no matter if the businessman and the engineer are "metaphorical archetypes," as Daniel Aaron argues, or if the book is in fact a joke, or if the work is a straightforwardly revolutionary tract as advertised. No matter how radical for his time, Veblen misunderstood the dominating nature of machine rationality. He fell into the same means/ends trap previously decried by Bourne, as Mumford saw: "What matters it if industrial society is run efficiently, if it is run only further into the same blind alley in which humanity finds itself today?" His debt to nineteenth-century utopianism propelled Veblen into a niche undeniably close to Comte and Bellamy. In short, Veblen, the champion of workmanship, elevated mechanistic modes of reasoning into a panacea. C. Wright Mills perceived that Veblen "failed to recognize the terrible ambiguity of rationality in modern man and his society." Even though he understood clearly the pecuniary nature of industrial society, Veblen still underestimated the power of capital; Theodor Adorno wrote that Veblen "dislikes capitalism for waste, not for exploitation." Like the other industrial utopians, Veblen took seriously neither democracy nor "irrational" aspiration. Playing essentially the same game as the social scientists, industrial managers, and municipal reformers, Veblen differentiated—and alienated—himself only by his idiosyncratic house rules.[30]

. . . .

For Veblen, Croly, Gay, and the other advocates of rational political rule, the war served symbolically to illuminate possibilities even as it showed the limits of the state of the art all too harshly. Observations on this situation reflected the ongoing need to believe in a civic version of the engineering method and a nagging sense of having been denied a fighting chance to prove its worth. Within a few strongholds, however, rational reform was pursued, refined, and redefined in the 1920s. Secretary of Commerce Herbert Hoover, American social scientists led by Wesley Mitchell and Charles Merriam, and new foundation administrators such as Beardsley Ruml, Raymond Fosdick, and Edmund E. Day aspired to a reconciliation of rationality and government. These men maintained the spirit of technocratic progress after many observers rested in the hope or despair that progressivism's time had passed.[31]

In contrast, the formerly enthusiastic and influential voice of rational reform belonging to Walter Lippmann changed tone after the war. "The period of hysteria which you saw last winter is definitely over," he wrote his mentor Graham Wallas in 1920. "It has been followed by a profound apathy. I can remember no time when the level of political discussion was so low." Whether such sentiments spurred or validated his antidemocratic preconceptions is open to question, but Lippmann clearly lost faith in the masses after the war. Groups, he came to believe, could not make logical or collective judgments. The challenge in *Public Opinion* to the myth of an informed citizenry led easily to the recruitment of social technicians. "The need for interposing some form of expertness between the private citizen and the vast environment in which he is entangled" inspired his updated program. This social expertise would come from a new source. Engineers, joined by "statisticians, accountants, auditors, industrial counselors," would be replaced by social scientists. Lippmann's new expert "will acquire his dignity and his strength when he has worked out his method," just as engineers had risen to importance. The role of the public evolved; they were to shift the burden of social choice onto the "responsible administrator."[32]

Lippmann returned to these themes in *The Phantom Public* (1925). Citizens, he wrote, "who are the spectators of action, cannot successfully intervene in a controversy on the merits of the case." His was a "theory which puts its trust chiefly in the individuals directly concerned"; it minimized "interference from ignorant and meddlesome outsiders." In essence, Lippmann argued for a politics of interest groups. Experts, by now an assumed fact of modern life, rarely received mention, but this pessimistic, if realistic, view of civic debate further removed technical capacity from public responsibility.[33]

American involvement in the war was probably too brief to permit reflective assessment of its effects. It did hasten the agreement of previously disparate professional and other groups on new terms for the state, bringing many social scientists, managerialists, and engineers together in networks that shaped the following decade. That their aspirations went largely unfulfilled underlines the importance of linguistic coherence for communities of belief. Machines exhibited efficiency, showed results, and responded to command and control. As immigration, urbanization, secularization, and the ascent of a consumer economy destroyed old bases for at least partial consensus, hopes in rational reform imposed an artificial unity, purpose, and security on a most disorderly political topography.

5

THE GREAT ENGINEER

The man to whom many looked for leadership in the 1920s apparently had all the right qualifications. Familiar with engineering, managerialism, and the quickly changing world of social science, he had shown active and effective, not hand-wringing, compassion in the war. Speaking the language of laissez-faire while understanding its limits, he encouraged business and philanthropy, success and conservation. Substance, if not style, radiated from the man; his integrity was beyond challenge. But Herbert Hoover, striving to pursue efficiency, opportunity, and justice in the terms he understood, found engineering modes of thought to be of little use in calming a panicked nation during the Depression. After encountering large-scale failure for essentially the first time, he retreated into sanctimony and blamed the victims. Method alone, no matter how perfect, failed him in crisis. Before that miserable period, though, Hoover developed a technocratic language, outlook, and program that redefined American politics.

After the gratifying and spectacular success of his Belgian and American food programs, Hoover found himself ideally situated for the postwar transformation of the reform impulse. In him the logical, orderly, effective methods of the engineer and the social concern of the reformer appeared to fuse; his worldwide reputation for humanitarianism was extraordinary. Observers who hoped to reproduce war organization in peacetime rallied to support a man Morris L. Cooke dubbed "the engineering method personified." John Maynard Keynes called Hoover "the only man who emerged from the ordeal of Paris [treaty negotiations] with an enhanced reputation." Louis Brandeis wrote Felix Frankfurter in 1920 that "I am 100% for him. [He combines] high public spirit with knowledge, ability, right-mindedness [and] organizing ability." As one contemporary later noted in the midst of Hoover's 1928

Fig. 5-1. Herbert Hoover, 1927 (Herbert Hoover Presidential Library)

presidential campaign, his "following comes from those who see those [social] functions as so complicated and intricate in the modern world that they demand an executive who is an efficiency expert and a super-technician." His mastery of administration lent Hoover substantial appeal.[1]

Had Hoover run in 1920 as a Democrat, he might have won and thus avoided the ignominy of the Great Depression that became his legacy. The

New Republic, for example, noted in that year that "the people who are for Hoover are people with their eyes on the facts, not on labels and doctrines. . . . They do care that the next President shall be a man who can choose men, conduct great affairs, and act on a trained estimate of the facts." But Hoover himself needed some label. The Democrats had a credible platform to match his ideals; in him they saw a candidate with no political debts, no hint of partisanship, impeccable technological and managerial credentials, and worldwide popularity. His style of Progressivism could reach out, from a distance, to voters disillusioned by the "soft" moralism associated with Woodrow Wilson and William Jennings Bryan. The ranks of Hoover's supporters included many luminaries who invested in their candidate the future of reform: Jane Addams, Ray Stannard Baker, Herbert Croly, Walter Lippmann, Franklin Roosevelt, and Ida Tarbell. On March 30, 1920, however, Hoover followed up on letters referring to himself as a Progressive Republican with a public declaration of his loyalties.[2]

With typical backhanded ambition, Hoover specifically denied seeking the nomination but allowed others to mount a campaign, because if "it is felt that the issues necessitate it and it is demanded of me, I can not refuse service." The semicandidate, "deeply and sincerely bored by the whole affair," according to one Hoover acquaintance, did not take to the stump, preferring instead to work on hunger relief and industrial projects. Popular enthusiasm could not overcome opposition within the Republican leadership, and Hoover dutifully backed Warren Harding's successful campaign. His reward was a choice of several cabinet positions, and Hoover decided to become secretary of commerce, despite the post's lowly reputation. His eight years there transformed the department into a prototype of modern bureaucratic government.[3]

. . . .

By 1921 Hoover had developed an original and timely social code in pursuit of world reconstruction and domestic progress. His ideology was blurry and imprecise; language held particularly private and often idiosyncratic meanings for Hoover, who maintained strict editorial control over all of his political material even though writing came very slowly. But the words he used to express his ideas in books, articles, and speeches are in some ways as important as the ideas themselves. A firm belief in logic, a distrust of political rhetoric, and a heartfelt desire to uplift the country characterize Hoover's prose. Its aesthetic shortcomings can be seen as products of the earnest, often stolid pursuit of accuracy that could, on occasion, be a strength. Five central but overlapping elements of Hoover's thought—individualism, service, efficiency, decentralization, and an apolitical vision of politics—help furnish an understanding of the era.[4]

Hoover devoted his first major book outside engineering, *American Individualism* (1922), to the concept that supported his entire ideology. Far from affirming laissez-faire absence of constraint, the book inextricably tied personal aspiration to social responsibility. Hoover's commitment to freedom within order attempted to move beyond both what he called the "claptrap of the French Revolution" and the "rugged individualism" some attributed to him. As a global manager of communications and finance, Hoover saw better than many the extreme interdependence of the modern world. At the same time, the American ethos touched him deeply. He contended, then, that government should assure individuals "liberty, justice, intellectual welfare, equality of opportunity, and stimulation to service."[5]

Although he moved away from participation in the Quaker faith in which he was raised, Hoover reworked several tenets from his boyhood religion into secular variants. The commitment to public service, a neo-Quaker ideal, became a second aspect of his thought. Hoover's adulthood passion for the service ideal no doubt also grew from small-town origins, irrespective of his religion; he wrote in his memoirs that the mutual aid of farmers "was social security itself." Above all, however, it was the experience of wartime—with its relative singleness of purpose, extraordinary coordination, and social cohesion—that confirmed Hoover's belief in the possibility and necessity of public service. He once wrote that "being a politician is a poor profession. Being a public servant is a noble one." The call to service that recurred throughout countless talks to business and professional groups always stressed an updated sense of noblesse oblige. Public-mindedness was for Hoover an enlightened response to a society's problems by its leaders, not an invocation of an irrational altruism, and he redefined classical republicanism to include technical experts in the Madisonian elite.[6]

Hoover connected faith in efficiency to a tenacious confidence in reason. Like the Taylor management reformers, he made the pursuit of both a broad increase in productivity and a similar decrease in waste priorities of astonishing social possibility. Because of its apparently unambiguous moral character, efficiency became an unassailable virtue. He argued that "the whole basis of national progress, of an increased standard of living, of better human relations, indeed of the advancement of civilization, depends on the continuous improvement of productivity." Metaphysics had no place in social discourse; things were for Hoover what they were. "Exactness," he wrote in his memoirs, "makes for truth and conscience." Qualitative judgments, fuzzy and imprecise, irritated him. He told one group of engineers that "there is nothing so much needed in our nation and our civilization today as a replacement of qualitative thought with quantitative thought." The state of such a society was easily determined: "The standard of living is the direct quotient of the amount of commodities and services that are available among the total population." Causal reasoning would reshape such a world according to laws social scientists could discern. Based on quantification and beyond the ambiguity of value judgments, efficiency fit well into this worldview.[7]

Hoover's belief in decentralized, local authority instead of wide intervention by the federal government, a fourth aspect of his thought, contradicted the movement toward centralization in pursuit of economies of scale. He drew on his own career as a manager of far-flung enterprises and saw no inherent obstacles to localized efficiency. During World War I, Hoover granted subordinates a great deal of autonomy, avoided closely defined organizational structures, and promoted his workers on the basis of knowledge and capability rather than hierarchical location. At the Commerce Department, he supported microeconomic coordination (via voluntary industry and professional associations) as a means toward macroeconomic control: managers of firms were supposed to make their individual choices with national goals in mind, thereby keeping the federal government out of economic planning. Hoover complimented the Associated Advertising Clubs on their "self-government in the greatest form of which democracy has yet given conception— that is self-government outside of government." In contrast to a Rousseauian notion of an elusive but definable common good, he insisted that "progress of the nation is simply the sum of local progress." Statist intervention created "a great lot of people who wish to lean on the Federal and State governments," but "our country was built by pushers, not leaners." Hoover had no use for large-scale public bureaucracies that he feared would cause the nation to "go backwards the moment we destroy the initiative of our people by constant extension of Federal authority."[8]

Finally, Hoover's intense distaste for politics manifested itself in this antipathy toward government. Indeed, he never referred to himself as a politician, nor did he use the word *politics* in any but negative connotations. For him it meant partisan infighting, favors and debts, and empty rhetoric. Since "neither politics nor litigation will build dams or canals," he sought to bring to government the same managerial aptitude that had made American business so prosperous. He wrote that "the political lobbyist and the ward politician thrived in a generation when his [*sic*] prototype had become extinct in business relations." This apolitical administrative stance contributed to Hoover's wide appeal. According to the *Review of Reviews*, his supporters believed that "Hoover the politician—in the unfavorable sense—does not exist." Why, then, did he run for the presidency? "In order that his executive gifts not be lost to the nation," the article continued, "it was essential that he, for the time necessary to be nominated and elected, bow to expediency." Rational reformers widely viewed politics as a relic from a premanagerial era. Morris Cooke wrote in 1928 that "if he is elected, as he probably will be, Herbert Hoover may show us how to shed politics in some large way in the demonstration of a democratic form of government." The *Review of Reviews* saw in Hoover "one whose early training in science, adapted as the tool of a genius for organizing, has remained with him, the antithesis of the unstable ways of politics."[9]

Hoover's comprehension of political economy, however logical, could not negate the irrational whims of political actors. He recalled his mission in his memoirs: "I was convinced that efficient, honest administration of the vast machine of the Federal government would appeal to all citizens. I have since learned that efficient government does not interest the people so much as dramatics." Nevertheless, he achieved notable peacetime reforms that increased administrative sophistication. Barry Karl concisely summed up Hoover's vision of a modern administrative state, "founded no longer upon the intuitions of neopopulist do-gooders whose well-meaning ignorance produced chaos," when he wrote that "the war to end war was now succeeded by the politics to end politics." With loftier goals than many Progressives, Hoover looked beyond merely nonpartisan government. Much like Frederick Taylor, who sought to supersede the profit system with a precisely quantitative measurement of industrial virtue, Hoover believed in an administrative method that could transcend politics itself. Not the first reformer to do so, he was probably the most powerful.[10]

．．．．．．．

Convinced absolutely of the primacy of logic in human affairs, Hoover sought to govern democratically through the pursuit of what might be called collective reason. Given his lifelong distaste for rhetorical flourish and appeals to ideology, he preferred calm discussion of the facts. He consistently attempted to formulate rational methods of cooperation that would decisively settle debate. Because the infusion of scientific rationality into engineering had been the source of its modern success, he wanted to bring the same spirit and method to politics. As he wrote in the press release accompanying the 1933 publication of *Recent Social Trends*, Hoover wanted to "project into the field of social thought the scientific need and the scientific method as correctives to an undiscriminating emotional approach." His first application of rational principles to social issues had come before he entered government. During his administration of men and material, he came to see the engineer as a "buffer" attenuating the characteristic antagonism between labor and capital. *Principles of Mining* contended that labor unions were the "proper antidote" to unlimited capitalist organization. Capital and labor, meeting as rational parties before impartial referees, could then come to agreement as they realized their mutual dependency. Hoover later demonstrated his commitment to this mode of negotiation by contributing the foreword to the American Arbitration Association's 1927 yearbook.[11]

Once he rose to the Commerce Department secretariat, Hoover faced the more difficult problem of reaching public consensus on wider issues. The engineer or arbitrator could help unions and capitalists see their common interests in the context of the business organization, but how was such rational discussion to proceed among citizens—and voters? As print advertising became more sophisticated in the 1920s and radio reached millions of homes quickly after being commercially introduced, Hoover adapted the strategy of intense publicity that had served him so well in his tenure as a food administrator in World War I. The flood of press releases issuing from the Commerce Department naturally kept Hoover's name before the potential electorate, but, more important, it sought to convince the public of the unarguable rightness of his conclusions. Rather than creating a public relations facade for the ruthless application of cold logic, he premised his use of publicity on reasonable conclusions that he believed would be seconded by the citizenry when he chose to consult it. Hoover's persona as an engineering mastermind, meanwhile, resulted less from conscious image making than from predictable responses of the nation's press at the height of the "machine age" discussion.[12]

Farther from the public eye, Hoover attempted to mobilize collective reason within what Ellis Hawley has termed an associative state. Parties with common interests came to Washington for meetings based closely on Theodore Roosevelt's White House conferences in order that they could jointly determine a rational course of action. In this way Hoover could maintain government involvement in the economy with little danger of being tarred with the antisocialist brush still dripping at the end of the Progressive Era; the laissez-faire ideal had left a legacy of profound antipathy toward centralized authority among business leaders. He became adept at bringing philanthropic foundations, social science researchers, and business leaders together in various combinations to exchange information while remaining true to his commitment to private initiative. In his first five and one-half years at the Commerce Department, Hoover sponsored 1,250 such conferences. As Hawley concludes, Hoover saw himself as antistatist, an advocate of positive government, and logically consistent. Avoiding the poisonous "interests" of "politics," he understood government as an administrative enterprise with reason as its primary tool.[13]

Before he implemented it on a wide scale in the 1920s, Hoover tried to instill associationalism among one of its most logical constituencies. After his successes in the war, Hoover bathed in the adulation of many engineers, who hoped that he would expand into peacetime the measures that had brought them added social influence. But professional unification, engineering reformers felt, had to precede effective social involvement. The Federated American Engineering Societies served as an umbrella organization that allowed civil, mining, electrical, and mechanical engineers to transcend professional boundaries and speak to the nation's needs with one voice. Hoover, the overwhelming choice to be the group's first president, accepted. After exacting dues and other support from constituent organizations proved to be impossible, the Great Engineer faced the irony of seeing his associative model fail among men heralded as society's most forward-looking professionals. Self-interest and mutual suspicion prevailed, the engineers' rational methods notwithstanding.[14]

Nevertheless, the ethic of engineering figured prominently in Hoover's most visible attempt to create a modern state. The need for a more efficient society led him to mount a large-scale campaign to reduce waste in industry through a variety of means. Expanded standardization of everything from screw threads to mattress sizes attracted wide public attention, and for a rapidly maturing industrial nation, increased interchangeability conferred substantial advantages. (One economic historian contends that "the significance of [the standardization] strand of Hooverism is perhaps better appreciated in countries that did not experience a similar technocratic intervention in the 1920s than it is in the United States." Present-day encounters with English electrical apparatus bear out this insight.) Hoover spoke on this theme repeatedly while at the Commerce Department, because standardization and waste reduction provided incontrovertible social benefits.[15]

The waste program's unambiguous gains stood in stark contrast to most moral choices where gray area predominates. After disclaiming that "there is no panacea for all of our economic troubles," Hoover articulated the attractiveness of such a pristine pursuit: "It is a certainty that the elimination of waste—and I speak of it not in the narrow sense but in the sense developed by all of our scientific and economic bureaus—is in itself an asset and not a liability." By the mid-1920s Hoover had even defined his terms narrowly enough to differentiate himself from the multitude of efficiency prophets. He insisted on "elimination of waste" as the chosen phrase for his program, stating that "I find difficulty in the use of the term 'efficiency' in this connection because that term has come to imply in the public mind a certain ruthless inhuman point of view." Hoover was determined to retain a humane and beneficial connotation for his pursuit, the archetype of the social solution advocated by a rational man.[16]

The strategy of collective reason most closely tied to Hoover's engineering training developed in the early 1920s and became a key aspect of his presidential policy. Rational public discourse, in his view, foundered in unpredictable currents of emotion and sentimentality. What was needed were facts: unemotional, objectively verifiable, and agreed upon. Probably unconsciously, Hoover relied on the same logic that had inspired the muckraking press, for if the details of a given social problem could be brought to public light, reactive corrective action would be inevitable. By 1920, however, the glory years of muckraking had passed, in large measure because the public did not react reasonably—well-publicized problems remained unsolved. Unlike Walter Lippmann, who doubted the public's capacity for rational reform, Hoover held fast: "If the facts can be established to an intelligent people such as ours, action is certain even if it is slow." Engineering trained men to ascertain and respect facts single-mindedly, making them most attractive in this vision of social discourse. Engineers, he said, "are selected on ability and precision of thought, men who are anchored to facts, whose whole background, mentally *and morally*, is fidelity to the determination of the fact, and dealing with the fact."[17]

Because few engineers endorsed associationalism—most directed their efforts toward technical challenges and not social reform—Hoover looked instead to social scientists for the assembly of social facts. Eager to improve their professional status after brief but promising involvement in government-sponsored research during the war, sociologists, economists, and political scientists operated under assumptions closely related to those of the engineers. These rational reformers sought to remedy the imprecision and emotion of American society with what they saw as their antithesis: science.

In the 1920s both parties found the timing perfect for such symbiosis.

Hoover wanted accurate data as a precursor to rational problem solving; upon his election to the presidency, he "felt that our first need was a competent survey of the facts in the social field." At the same time social scientists gained experience and exposure. As Wesley Mitchell had said to a meeting of statisticians after the war, "in short, the social sciences are still childish." Their wartime experience had shown Mitchell, Edwin Gay, Felix Frankfurter, and others that social planning on a national scale necessitated quantitative skill beyond the current state of the art. By working in governmental and quasi-governmental research projects, social scientists could accumulate sufficient data and methodological capabilities to gain credibility with universities, funding agencies, and, often, policymakers.[18]

Hoover himself became the foremost audience for these investigators. He found the social scientists he met during the war neither political appointees nor profiteers, but idealistic outsiders motivated by professional opportunity and a sense of public service. By including them within his buildup of semipublic networks of coordination in a number of capacities, he could begin a program of national economic planning while compromising neither his Republican party identity nor his social ideals. The 1921 unemployment conference, an early step toward this associational model, led to the production of a two-volume survey of recent economic changes. Hoover got facts, and the economists got the attention of a political administrator. He also invited statisticians and economists to complete a survey of current business. Much of the government's basic economic data remains a reminder of Hoover's insistence on accurate figures. Deeper original involvement of social scientists in national government began in 1929.[19]

Within six months of his inauguration, Hoover appointed the President's Research Committee on Social Trends. Told to provide him with the data upon which to build a rationally good society, the body responded to his "desire to have a complete, impartial examination of the facts." The committee included Howard Odum of the University of North Carolina, William Ogburn and Charles Merriam of the University of Chicago, Alice Hamilton of Harvard Medical School (added on Hoover's request that a woman be involved), and Columbia's Mitchell, who had also served on the earlier panel that produced the survey of Recent Economic Changes. Funds for the massive compilation of social data came from private sources—the Rockefeller Foundation—as they had for many of Hoover's projects under Harding and Coolidge. A network of former war bureaucrats continued to help Hoover organize philanthropic support.[20]

Hopes for the investigation of Recent Social Trends ran high. Traditional issues, in Hoover's plan, "could be viewed through a rational, scientifically organized vantage point for the first time." Emotion, characteristic of non-Anglo-Saxon peoples and of women, would be rendered politically irrelevant. His confidence in social science's engineeringlike method led Hoover to see his presidency as the moment when scientific progressive reform would be refined and vindicated. The Depression dashed that hope, and *Recent Social Trends* was not released until after the 1932 election: some authors wanted to avoid the appearance of political partiality, and most of the committee differed sharply with Republican Depression strategy. The timing of the report prevented him from developing a practical program based upon social facts. Hoover's ideal of collective reason gave way to more centralization of power and expertise in the same federal government that he had tried both to use and to restrain.[21]

. . . .

Partly because of the example of Hoover, rational reformers looked to a metaphoric—vague and imprecise—version of engineering for political regeneration. Such language had a powerful appeal, especially given the era's many and rapid advancements in technology. John R. Dunlap wrote in *Industrial Management* in 1921 that "we can now easily solve each and every problem that confronts the American people, provided we follow the advice of a trained engineer who knows precisely what he is talking about." The Engineers' National Hoover Committee, a nonpartisan campaign group, told its members that "there has probably been too much said of Hoover as an engineer, but there never can be too much said of the engineering method in modern public affairs as opposed to the method of talkative and disruptive radicalism." Hoover himself praised the merits of this approach to politics: "From the point of view of accuracy and intellectual honesty, the more men of engineering background who become public officials the better for representative government." Hoover also wrote elsewhere that "our political and economic problems call for the application not of any set doctrine, or fixed formula, or principle of deduction, but of the scientific, inductive method." Individuals, then, mattered less than the technique they employed; engineers were factually honest, administratively effective, and did not talk too much.[22]

Hoover became a hero despite the anonymity implied by his statements. His managerial successes during the war and an exciting climate of technological innovation made the engineer-cum-administrator label a most favorable one. The usual political attributes—charisma, leadership, vision, experience—faded as administrative capabilities came to the fore. One reporter told the nation that "to talk to him socially is to approach a stone wall, a vague, gentle, absent-minded stone wall. . . . He listens, he looks at one thoughtfully; and meanwhile his mind is ranging far away. He is estimating the capacity of a cattle ship, the justice of a miner's working day and wage, or the question of primary-school lunches as related to small primary-school tummies." The supposed authority of engineering turned a lack of political and communications skills into a political advantage.[23]

Already known in 1903 as the Great Engineer at age twenty-nine, Hoover continued to be stuck with the label during his political career. His close friend Mark Sullivan wrote that he "regarded our entire business structure as a single factory, conceiving himself, as it were, consulting engineer for the whole enterprise." Hoover accordingly "set about applying to the whole business structure of the United States principles similar to those which Henry Ford applied to the manufacture of automobiles." John Dunlap insisted that "it is simple truth to say that Herbert Hoover knows more about how to solve every economic problem that confronts the American people than any other man in public life." Dunlap's logic for such a judgment was as rigid as steel: "The solution of these problems is second nature to Herbert Hoover—because he is an engineer, and because each and all our difficulties grow out of industry and hence they are engineering, pure and simple." When Hoover ran for reelection, Zay Jeffries of the Engineers' National Hoover Committee pleaded for support because "the whole engineering profession and engineering methods are on trial." Though Jeffries argued that "we may expect soon to see order come out of what some may have regarded as chaos," few others attempted to muster support with engineering images during 1932.[24]

After he left the field, Hoover seldom referred to himself as an engineer. Moreover, he steadfastly resisted efforts in the press to tie his administrative style to any engineering practice, no matter how positive the public image of the profession. In a remarkable letter to the *Chicago Daily News*, Hoover, who almost never complained in print about journalistic treatment, explicitly denied the newspaper's contention that he "wanted to apply the exactness of engineering science to the problem of hard times" during the recession of 1921–22. "I have never made a suggestion of this character in any shape or form, nor have I ever had a conference of that import or published a statement of such character," he protested. Why such an explicit denial of everything his supporters were saying about him? Hoover actively avoided the crude mechanism typical of the early Taylorites and, later, the Technocracy movement. He quite likely did not see himself as an engineer in politics, but rather as a public servant with an administrative agenda, and he resisted the characterization even while earning it. Nevertheless, Hoover repeatedly described the nation, explicitly or metaphorically, as a machine or a factory throughout his letters and public statements. His use and understanding of technical language illustrate the persistence of engineering modes of thought in the political discourse of the 1920s.[25]

Hoover made his reputation as a problem solver, and his rhetoric frequently framed social and governmental projects in terms familiar to engineers. The Great War had exposed "the weaknesses in our legislative machinery"; legislatures were, after all, political in the negative sense of the word. Finite social problems, however, would yield to the engineering method, the "antithesis of politics," for "*there is somewhere to be found* a plan of individualism and associational activities that will preserve the initiative, the inventiveness . . . of man and yet will enable us to socially and economically synchronize this gigantic machine that we have built out of the applied sciences." Nobody in America "could make a better contribution to this than the engineer." A Platonic ideal of a voluntarily coordinated state led Hoover to favor industrial technique and its analogues as resources for political renewal.[26]

The language of engineering continued to emanate from Hoover's office during the 1920s, and the premises he and other rational reformers brought to American politics originated in the machine process. Using a mixed metaphor not uncommon in the period, Hoover charged a committee "to visualize the nation as a single industrial organism and to examine its efficiency toward its only real objective, the maximum production." Such an inquiry was especially timely, said Hoover, because "our national machine is today doing worse than usual, as witness the 3,000,000 idle men walking the streets" during the recession of the early 1920s. Examining producers only began the process of social evaluation; advertisers needed to be careful of creating unrealistic desires if they were "to maintain a position as part of the economic machinery of the country." Hoover's mechanistic conception grew broader still. Differentiating between political and economic structures, he told a Stanford graduating class that neither America's "form of government nor this vast machine of production and distribution" could withstand the shocks of class conflict. Even the household could be described in the language of production: "The home is the family workshop," read an address to the General Federation of Women's Clubs. "The equipment and organization are an index of its efficiency."[27]

But the machine metaphor included more than the nation's many economic relations. No great social objective could be attained "without painstaking analysis of the facts and forces," after which "open minds" must be "willing to hammer every proposal on the anvil of sincere debate unmixed with debasing alloys of malice and selfishness." In the consideration of these questions, "we must go far deeper than the superficials of our political and economic structure, for these are but the products of our social philosophy—the machinery of our social system." America's social code was not at base democratic but individualistic: "Democracy is merely the mechanism which individualism invented as a device that would carry on the necessary political work of its social organization." Mechanism, invention, device, work, organization—this is the vernacular of engineering, of controlled cause and effect. Hoover spoke and wrote in the language of his professional training as he defined his conception of political economy and even the fundamental ideologies of his nation. As we have seen, these expressions were not the rhetorical ornamentation of a politician seeking to energize a crowd; such was not his style. He did realize how technical language could begin to redefine politics.[28]

Despite the imprecision of technical terms in popular discourse—"I have come to wonder whether some people know the difference between a cow and a kilowatt"—Hoover praised the infiltration of such language into common usage: "The fact that the public is beginning to use terms which have even an indeterminate background is itself evidence that the engineering view is at least in a small way penetrating the public mind." In the same address he set out the larger project: "If our engineers would be a little more vocal . . . on all public questions . . . we could transform the thought of this nation within another twenty years." His use of technological metaphor, then, represented one aspect of a calculated attempt to reshape discourse in order to transcend conventional politics.[29]

"Primarily, social life is organic and not mechanical," Hoover wrote in one of his most fascinating documents, the foreword to Elisha Friedman's *America and the New Era*, published in 1920 when his understanding of society was taking shape. The statement's contradictory appearance at one level demonstrates Hoover's human inconsistencies; reflectiveness was not a strong suit, wordplay not a pleasant diversion. While showing that Hoover, like the rest of these reformers, spoke in a variety of dissimilar political languages, the statement does not undermine his later use of frequent mechanical metaphors. Instead, the disclaimer can be seen to refute Marx's claim to a science of society. Inflexible social laws could not prescribe the shape of politics; such an assertion directly contradicted Hoover's commitment to individualism. A "pure" science of society would never do, but an *applied* science of society, much more pragmatic and superficially less ideological, solved problems without imposing political strictures upon the populace. Engineering's objectivity appealed to noble instincts, leaving society to an organic, individualist determination of its fate. Hoover wrote later in the same foreword that "terms must not be confused with realities, or labels with conditions. We must face concrete facts, rather than attempt to apply doctrinaire generalizations." Dogma resided in the province of ideology; Hoover sought to define a politics of technique.[30]

This attempt met formidable opposition when the Depression refused to yield to the most sophisticated and rational instruments in Hoover's intellectual toolbox. His sizable intellectual investment in mechanical metaphor also foreclosed consideration of other approaches to a crisis of spirit as well as of finance. In contrast to Hoover's position as the butt of cruel jokes and historical disparagement, however, his social scientists fared better. They expanded the influence they had begun to gain under Hoover in the Roosevelt administrations. For the most part, they continued to practice the antipolitics of rationality with political "cover," just as Hoover had done at the Commerce Department. The difference came when he inhabited the White House and caught the attention and abuse directed, for the most part, elsewhere when he served under an electoral figurehead. Advocating government through technique, Hoover fell victim to traditional politics, while many of the investigators he so fundamentally aided went on to work for a Democratic successor.

....

The historian William Appleman Williams suggested that Hoover was abandoned by the people his vision sought to enable; the Great Engineer had dreamed "that the people—the farmers, the workers, the businessmen, and the politicians—would pull themselves together and then join together to meet their needs and fulfill their potential by honoring the principles of the system." This is too easy. Hoover may have been tragic in that his flaws grew out of supposed virtues, but the attempt to prescribe to the nation a capricious mixture of altruism, objectivity, and individualism based largely on his own example suggests a fall based on hubris rather than desertion. Walter Lippmann wrote Felix Frankfurter in 1930 that "when men of his temperament get to his age without ever having had real opposition, and then meet it in its most drastic form, it's quite dangerous." Dangerous that confrontation was, to Hoover and to the nation.[31]

The ideal of service, no matter how laudable, embodied substantial inconsistencies. As his failure to interest engineers in the associational approach demonstrated, Hoover only intermittently persuaded the people who praised individualism to forgo private advantage for a larger good. In an address to the American Engineering Council, he said that "no engineer can receive any material benefit from [this association]. It can advance no economic interest." Why, then, should anyone have participated? In another speech, he reiterated the same theme but contradicted himself in his attempt to convince. After claiming that the engineer is "disinterested," Hoover announced that "we are advancing the interest of the engineer as it has never been advanced before." Hoover gambled by attempting to motivate altruism by appealing to self-interest, but he had seen the service ideal work. To Howard Heinz, the pickle magnate long praised for his humane labor practices, he wrote, "If we had this sense of responsibility throughout the whole industrial world there would be no industrial problem today." Heinz and Hoover, though, set a lonely precedent for their peers.[32]

Hoover's idealized memory of the war continued to serve as a mountaintop vision. He fondly recalled its "vast sense of national service and willingness to sacrifice" but neglected to reflect on the clumsy coercion and the resulting mobocracy engendered with the sacrifice. In the prosperous 1920s he saw "more need for this unselfish devotion . . . than even in the war." To his credit, Hoover put before America a noble ideal, however naively. He sounded "not a call for the high emotion and glamor of the war but a call for citizenship based upon the daily obligations to the community and not upon the privilege to dominate or exploit it." The need for a sense of service and his hopes for enlightened civic responsibility supported the rest of Hoover's social philosophy. "Associational activity can be made a safeguard to the individual and to equality of opportunity," he told one graduating class. But without "the devotion to service, we cannot hope to escape being crushed by this gigantic machine which we have created."[33]

Even deeper than the service ideal lay Hoover's unquestioned faith in rational method. No issue, even a moral one, could withstand the march of reason: "The line between right and wrong is difficult to draw, and the national common sense is the instrument that will ultimately determine it." The engineers who served in the war "vindicated the scientific attitude in dealing with problems of social organization. Unknown difficulties succumb to scientific analysis." Hoover shared that vindicated method, so his judgments stood, Olympian, beyond the push and pull of conventional politics: "Any intelligent person who has the patience to read and think these problems through, and to familiarize himself with the methods we have developed for their correction, will find that these efforts are in the interest of the public." This attitude reinforces the historian Carl Degler's assessment: Hoover was not above politics, as he claimed, but, confident in the power of the engineering method, merely resistant to compromise.[34]

In a letter to Herbert Croly in 1930, Walter Lippmann offered a glimpse of the dilemma facing Hoover's supporters. Despite his shortcomings, Hoover still represented much of what the rational reformers sought to accomplish. Lippmann could not "quite bring [himself] to condemn him completely," for "underneath all his failures, there is a disposition in this Administration to rely on intelligence to a greater degree than at any other time, I suppose, since Roosevelt." For a decade Lippmann had been trying less and less successfully to hold irrational citizens and intelligent administration in theoretical tension as Hoover tried, by managerial means, to steamroller one with the other. Hoover gained fame at the Paris peace talks, for example, partially because he had no constituency to consider. Without electoral responsibilities, he could allow pure rationality free play. But Hoover attributed his insights not to his fortunate position but to his method. It was, after all, a heady time for the engineer. "Technology and science were steadily lifting mankind toward more beauty, understanding, and inspiration," Hoover recalled. To ignore the possibilities for a science of reform and a method of government would be irresponsible, for as technology had defined society and its problems, so should it, in this argument, realign politics.[35]

The presumed objectivity of the engineer and social scientist implicitly worked in a capitalist framework, taking it as a given and not as a result of political processes. The question of choosing ends for these marvelous means was never raised; finding and implementing the method sufficed in Hoover's understanding of politics as technique. Even some of his supporters expressed reservations about Hoover's thoroughgoing faith in method. In his review of *American Individualism* in the strongly pro-Hoover *New Republic*, the philosopher Morris Cohen at once shared some of Hoover's hopes and presciently forecast his shortcomings. Cohen began by outlining a social role for the engineer possibly even broader than the one Hoover encouraged the profession to adopt. "Though technical ability alone can never elevate a man to supreme leadership," he wrote, "as the engineering profession expands, more of its members will find their training a great advantage in the struggle for rulership of human affairs." Disputing the book's assertions about politics—"Hoover intimates that it is only demagogues who feed the mob with emotional phrases"—Cohen found Hoover guilty of the sins he condemned: "As an engineer he surely could not get very far with such loose, unverified statements."[36]

The reviewer identified terrific flaws in Hoover's attempt at social philosophy before denying the notion of politics as engineering any validity: "Human nature is still a factor in human affairs and the training of the engineer has not as yet, despite cheap cant about social psychology and motivation, prepared the engineer to replace those who, with less knowledge of the material facts, have a better prophetic sense of what will go with multitudes of men." Cohen grasped what Hoover could not: humans are not merely rational creatures, leadership cannot be reduced to method, and truth is more than collected facts.[37]

Hoover continued to base his politics on a model of reliable causation even after his defeat in 1932. For the engineer, ambiguity and compromise, the stuff of politics, were anathema. Hoover's understanding of political language as fixed, finite, and final informed a 1934 letter to Wesley Mitchell. Hoover resisted Mitchell's advocacy of a centrally planned economy, saying that "the New Dealers appropriated a perfectly sound term [planning] and assigned it as a meaning and a justification of Regimentation of Socialist and Fascist qualities." Instead, Hoover attempted to redefine political language and neutralize the New Deal program: "The whole scheme of democratic government is an attempt by Parliaments, Congresses, Executives, and Courts to include 'National Planning.'" Mitchell's problem was "a lack of faith in these organisms." Colored by rage, Hoover's view of the New Deal overlooked his proximity to its origins.[38]

Struggles over political meanings exasperated Hoover. He asked Mitchell, perhaps in cynical jest, for impossible definitions of the contested terms of political debate:

Perhaps as much as anything at the moment, we need that Wesley C. Mitchell, acting as umpire, shall give us a workable definition of the following terms:
Liberalism
Radicalism
Conservatism
Reaction
National Planning

Socialism

Fascism

Economy of Abundance

Economy of Plenty

Capitalism

The Competitive System

The Profit System

Laissez-faire

Individualism

And a dozen more. We could then refer to the Mitchell standard and save a wealth of ink and discussion and establish some clarity of thought —and convict a lot of liars.

The notion that democratic politics consists in precisely what he sought to curtail—a "wealth of discussion"—escaped Hoover and the rest of the technocratic progressives who paternalistically sought "clarity of thought" for the good of the state's merely human citizens. The irony is that Hoover could be hydraulically fluid in his own definitions, often "uniting opposites by proclamation," in the words of one biographer.[39]

In his social philosophy, his political language, and his networks of public and private power, Herbert Hoover undertook a self-conscious program of rational reform. His mining experiences, wartime prestige, and administrative capacities provided him with both the motivation and the methods to attempt a reconfiguration of American politics along engineering lines. After individual voluntarism and collective reason failed to remedy the Depression, he refused to entertain the possibility that his methods and assumptions, rather than the American people and European economies, were inadequate. The same economic disaster, however, spurred both intensified efforts to engineer society and, later, increasing dubiety as to the ultimate capabilities of social engineering.

6

SCIENTIFIC PHILANTHROPY, PHILANTHROPIC SCIENCE

While the hothouse reform climate of the 1910s cooled somewhat after the Great War, in the 1920s well-situated philanthropic managers and social scientific entrepreneurs developed a program of rational change in less public venues. During the interwar period the interconnections between social science and big philanthropy influenced both the intellectual content and, more important for our purposes, the institutional structure of social research. A small group of administrators and investigators—many of whom served in the war or trained in the rapidly modernizing American research universities—drew together after the war. They then erected an extensive infrastructure of economic and political research and application based on a literal analogy between social investigation and natural science and engineering. Intellectual, financial, and institutional commitments to scientistic reform increased dramatically as a result.

. . . .

Theoretical support for new programs and approaches originated outside the core social sciences. Until psychoanalytic perspectives captured American momentum after 1930, some rational reformers appropriated behaviorist psychology to promise the conquest of the irrational. Unlike Herbert Croly and the Taylorite engineers, who had argued that humankind could shape its environment, behaviorists sought to make people worthy of existing institutions. Asserting that "there is not the slightest iota of choice allowed to any individual in any act or thought from birth to the grave," the social scientific historian Harry Elmer Barnes paternalistically sought to give people "a better set of experiences through heredity, education and association [determined by] natural and social scientists." Adaptation, normality, and conformity became primary social values. Knight Dunlap, a sociologist, declared in 1920 that "the scientific psychology—empirical . . . [and] logical—alone offers any chance of finding the help which society needs." Dunlap and others wanted to rationalize societies at the individual, not collective, level.[1]

Elitism comprised an essential element of the new "differential psychology" that, wrote Barnes, "has given scientific confirmation to the old Aristotelian dogma that some men were born to rule and others to serve." E. L. Thorndike was blunter still. "The argument for democracy," read an article in *Harper's*, "is not that it gives power to all men without distinction, but that it gives greater freedom for ability and character to attain power." By 1930 psychology as the rectification of the irrational informed much of American social thought. An important avenue for its insurgency was provided by the increasingly "scientific" social sciences when a rigid positivism came to bear on human societies. As one political scientist wrote in 1923, "The scientific mind . . . does not limit its faith in cause and effect to the physical world, but extends it to include man; it regards man as a part of the mechanistic universe; it rejects the doctrine of freedom of the will as incompatible with the scientific attitude." An explicitly antidemocratic ideology of professionalization and progress quickly took shape, with "science" at its core.[2]

Scientism penetrated political science, sociology, and, less completely, economics in the 1920s, as the intellectual historian Dorothy Ross has shown. Wesley Mitchell and Charles Merriam, later two-thirds of the New Deal's National Planning Board, led the movement to transform mere good intentions into social engineering in their academic and popular writing as well as in their organizational acuity. Though the Depression brought some of their methods and conclusions under attack, these men set much of the tone for their respective disciplines over nearly a quarter-century. In the same period, social science achieved some of the professional solidification that had characterized engineering in the first two decades of the century. Fields of investigation less than a half-century old developed a nearly obsessive fascination with their own history by honoring past "masters" of the field—some still living—in rhetoric and discarding them in practice. Social scientists' concern with their own history coincided with ever more determined efforts to free their disciplines from historical counterparts—political science from history, sociology from ethics, and economics from political economy. Academics escaped from history by relying more heavily on "science." Theorists of the founding generation—people like Veblen, Albion Small, Charles Horton Cooley, and William Graham Sumner—gave way to a new breed of successor committed to objectivity, quantification, and methodological elegance. In each field, however, the ahistorical scientific outlook took root with different degrees of tenacity.[3]

Political scientists oriented themselves toward the pole star of scientific control in large numbers. James T. Shotwell of Columbia, a veteran of Woodrow Wilson's Inquiry group, counterpoised "emotions, prejudice and ignorance" to "knowledge under control," pleading that "we must learn to deal with social facts as Watt dealt with steam." Was not American democracy more important than the light bulb? "It never seems to occur to one," wrote Shotwell, "that there is anything strange in the fact that we have better equipment for studying electricity than for studying society." Political scientists stood where the engineers had in the late nineteenth century, searching for a method capable of generating answers adequate to the task before them. In an epoch-defining address to the APSA, Charles Merriam confidently predicted "that we may definitely and measurably advance the comprehensiveness and accuracy of our observation of political phenomena," and that the mechanisms "of social and political control may be found to be much more susceptible to human adaptation and reorganization than they now are." Applied science allowed Merriam to hope for a reinvigorated version of democracy.[4]

Behaviorism persuaded many sociologists that science could make social control, the longtime aspiration of social scientists dating back to Comte, a live option. George Lundberg promised in 1929 the birth of "exact social sciences and a consequent transformation of the social world comparable to that which the physical sciences have wrought in the physical world." Similarly, Barnes applauded a modern society that depended "upon social science for adequate and intelligent control, direction, and reorganization." Franklin H. Giddings, one of the most forceful of these behaviorists, drew inspiration from their similarities with engineering practice and turned an intense bitterness engendered by the war into a fervent pursuit of scientific sociology. The engineer—contrasted with "untrained and unchastened uplifters" and the "phosphorescent ignorati of revolt and revolution"—was called to address "a world full of nations crippled and exhausted by a devastating war." While social engineering involved "an acceptance of scientific principles as the basis of practice, and a following of technical methods in applying them," the potential benefits remained untapped. "It is hardly necessary to insist," insisted Giddings, "that the engineering way has not been followed extensively in social reform, class struggle, public policy, or legislation."[5]

Within academic economics, Mitchell and a small group of like-minded innovators labored on several fronts to "prevent economists from running away from their essential task—the task of understanding and guiding our economic life," in the words of one practitioner. Knowledge, in this view, was justified in its application. Mitchell wrote in 1934 that the "hope of building up a social and economic system . . . in the complex world of today rests upon the progress of the social sciences quite as truly as our hope of controlling natural forces rests upon the progress of the physical sciences." Similarly, Jacob Hollander had told the AEA in his presidential address of the year before that economists could "prevail upon affairs" by using scientific methods and refraining from nineteenth-century moralism. With such a program, they could "be fearless in the knowledge that is power." Nevertheless, the dominant neoclassical model of the American economy retained supremacy, and the institutionalist challenge of Mitchell and his colleagues like Walton Hamilton never changed the mainstream orientation of the profession.[6]

All three disciplines' theoretical visions of the scientifically mandated state shared a fascination with process and method. Painfully visible ethnic, economic, and geographic facets of American pluralism no doubt influenced attempts to find a methodological, rather than substantive, binding force. The moralistic heritage of nineteenth-century religious rationalism also carried its stigma. In the 1920s, however, borrowing from science and engineering held possibilities not only for social progress but for professional importance. The banner of science enhanced a facade of social disinterestedness and increased researchers' credibility as agents for the status quo.

. . . .

Social scientists and political executives often reconciled advocacy and objectivity in semiofficial—and largely unaccountable—relationships; institutional developments accompanied evolving intellectual positions. Journals multiplied, professional memberships increased, and operating budgets escalated. Quantitative research designs expanded in scope and complexity, with such centers as Columbia and the University of Chicago in the lead. Graduate programs grew and provided further assistance to large-scale social investigation. The SSRC, founded in 1923, partially centralized the distribution of funding and the coordination of research. According to one account of the enterprise, it sought cooperation between "students of politics and the other branches of social science and also with the students of psychology, anthropology, geography, biological science, and engineering" in order that "the new political science may avail itself of all the results of modern thought in the attempt to work out scientific methods of political control." Extra-university centers for social, political, and economic research became more common in the 1920s as the IGR was followed by enterprises like the NBER. Large-scale and labor-intensive investigation, however, necessitated commensurately substantial funding.[7]

In the decades after World War I, American philanthropic foundations enabled the development of scientific social reform. After wealthy industrialists and merchants nationalized and rationalized their business enterprises in the late nineteenth and early twentieth centuries, they saw social problems still being attacked locally and on moral grounds. To counteract these apparently ineffective tendencies, Carnegie and Rockefeller began the practice of endowing foundations that applied to "charity work" the same principles—including economies of scale—that had made business enterprises so successful. In a development paralleling the evolution of the modern corporation, a new generation of professional managers steered the course of philanthropic donation, separating benefaction from control of the foundations.

New values accompanied this transition. Efficiency, frequently invoked as a political criterion in the 1910s, implied a model of inquiry and action based on engineering. Accordingly, many reformers viewed science as a replacement for politics, pejoratively defined. Coolheaded and rational, the scientist of society could succeed where deal cutting and emotional appeals to the lowest common denominator had failed. In this view, local churches and other centers of aid and comfort also carried the inefficient—and politically volatile—freight of ethnicity, morality, and autonomy. The modern solution to poverty, unemployment, and crime appeared to be efficient, not fair, and correct, not just, because the reformers understood their task in primarily managerial terms. Like Frederick W. Taylor, the social engineers used science to negate traditional moralistic concerns with equally moralistic but superficially apolitical objectives.[8]

The drive to power of philanthropic managers changed applied scientific thinking. Frederick Keppel, the chairman of the Carnegie Corporation who tended to steer his enterprise along intuitive and unsystematic lines in contrast to his Rockefeller colleagues, wrote in 1936 that the foundations "have, I fear, been the chief offenders in forcing the techniques of research which developed in the natural sciences" onto social science. In their encouragement and validation of a literal analogy between what Merriam called "natural and unnatural sciences," three parties—foundation administrators, social science entrepreneurs, and government officials—mutually encouraged a positivist and often mechanistic approach to social order. Of these men, Herbert Hoover worked from within government while Mitchell and Merriam led the way in social science. The key foundation operatives— Beardsley Ruml, Raymond Fosdick, and Edmund E. Day, for example—may have been less public, but their role in social engineering cannot be underestimated.[9]

These newly formed philanthropies began with relatively conventional concerns for social welfare, as expressed in the Russell Sage Foundation's support of social work. After the experience of World War I convinced many social scientists of the utility of applied scientific inquiry, however, the Rockefeller philanthropies led the search for systemic causes rather than for local effects. "The best philanthropy," John D. Rockefeller, Sr., had declared, "involves a search for cause, an attempt to cure evils at their source." This policy had governed the Rockefeller medical efforts from the outset, and it came to characterize the foundation's patronage of social research as well. The tendency to seek overall causes rather than to address symptomatic needs also served to remove social workers—who tended more often to be female, radical, and sympathetic to ethnic distinctiveness—from the process of policy formation. The foundations and the enterprises most suited to systemic social engineering were centralized institutions led by white male elites. By 1930 social workers, in particular, had substantially less contact with the architects of scientific social reform, although two exceptions—Sydnor Walker at the Rockefeller Foundation and Mary van Kleeck at Russell Sage—stressed their profession's worth.[10]

Various organizations had pursued scientific social reform since 1900. The American Association for Labor Legislation, for example, worked under the motto "Conservation of Human Resources" to mediate industrial strife. Such important social scientific advocates as Wesley Mitchell, Leo Wolman, George Soule, and John R. Commons of the University of Wisconsin worked with the group, to which Herbert Hoover looked for support in his efforts to relieve unemployment in the early 1920s. Another member of the association, E. A. Filene of the Boston department store, teamed with Henry Dennison, a devoted Taylorite, to begin the Cooperative League in 1919. The league became the Twentieth Century Fund in 1922 and supported managerial reform (with grants to the Taylor Society and the International Management Institute, where van Kleeck occupied an important position), fact-finding research, and publicity for social projects via Paul U. Kellogg's *Survey Graphic* magazine. The fund later initiated its own research projects, which remain ongoing.[11]

The first national policy group, the IGR in Washington, expanded to include an institute of economics in 1922 and a year later added a graduate school of economics and government. In 1928 the Brookings Institution, as the IGR came to be called, dropped the graduate school and focused on economic and political research. The IGR's statement of purpose cited as its primary supporters "men who believe that there should be a non-partisan, independent institution" to investigate and publicize "the most scientific practical principles and procedures that should obtain in the conduct of public affairs." In an even bolder, if more confusing, declaration of social engineering aspiration, the director of the Brookings Institution wrote that "it is believed that the most significant task of the present century is to control the new industrial civilization, to make the power-controlling sciences direct the power-creating sciences."[12]

Wesley Mitchell and Edwin Gay, meanwhile, assembled an eclectic board of directors at the NBER. Their goal was to insure credibility: with socialists and capitalists, workers and sellers overseeing its publications, the bureau's findings gained a reputation for authoritativeness. Unions referred to the economists' judgments in arbitration hearings while bankers and executives looked to the NBER for basic data on business cycles, wages, prices, and other fundamental aspects of the economy. The very usefulness of the NBER data, however, placed intense time pressure on the researchers, who sought to work more carefully with long-term trends. Reconciling timeliness and usefulness with reflection and validity continued to concern Mitchell and his associates at the bureau. Even so, more than the IGR or the still-young university research institutes, the NBER quickly became an important part of Secretary of Commerce Hoover's semipublic economic coordination program beginning with the Unemployment Conference of 1921. On his advice the Carnegie Corporation handsomely supported the NBER in the early years of its existence—as did the Rockefeller philanthropies—but after 1923 Frederick Keppel shifted the Carnegie philanthropy's focus away from large-scale social research.

In contrast to the skeptical Keppel stood Raymond Fosdick (1883–1972), a trustee and later president of the Rockefeller Foundation. The younger brother of the famous evangelist Harry Emerson Fosdick, he had served in the World War I mobilization, working on an almost daily basis with Keppel,

Fig. 6-1. Raymond Fosdick (Courtesy of Rockefeller Archive Center, Pocantico Hills, N.Y.)

and found traditional suasion inadequate to the task of warfare in particular and reform in general. His position at the Rockefeller Foundation made him a powerful advocate of social engineering. He told one graduating class in 1922 that "now as never before we need creative intelligence—knowledge consciously applied to our problem—the same kind of fearless engineering in the social field that in the realm of physical science has pushed out so widely the boundaries of human understanding." Fosdick acted on this conviction throughout his long career in the Rockefeller circle.[13]

With Fosdick's help, the Laura Spelman Rockefeller Memorial, a philanthropy originally intended to support churches and other conventional agencies of social amelioration, became the most active institution in the advancement of scientific reform. Under the direction of Beardsley Ruml (1894–1960) in the 1920s, the LSRM embarked on a terrifically expensive and wide-ranging program of thoroughgoing support for rational reform. Ruml, a University of Chicago psychometrician who worked with E. L. Thorndike in World War I, was an enigmatic administrator who worked under James R. Angell at the Carnegie Corporation. When Angell left to become president of Yale, he apparently convinced the LSRM board that Ruml would steer a steady course. Instead, soon after being chosen to head the memorial at the age of twenty-five, Ruml instituted a program of "social science and social technology." To justify the change in direction, Ruml argued that the memorial's "humanitarian" interest in social science was founded on "a belief that knowledge and understanding of the natural forces that are manifested in the behavior of people and of things will result concretely in the improvement of conditions of life." If Mrs. Rockefeller were still alive, the report seemed to say, she would be pleased that her legacy could help more people, at a more fundamental level, because the new LSRM invoked the model of engineering to bolster its claims.[14]

Once he had freed the memorial from a complete commitment to localized and evangelical social aid, Ruml worked to implement the methods and spirit of applied science in the service of ill-defined social objectives. An internal memo defining the program employed revealing terminology to lament that "all who work toward the general end of social welfare are embarrassed by the lack of knowledge which the social sciences must provide. It is as though engineers were at work without an adequate development of physics or chemistry." Because of the power of their methods rather than the wisdom of their objectives, Ruml endorsed the knowledge of experts as being superior to the desires of the many.[15]

Accordingly, the LSRM supported programs aimed at educating the mass-

Fig. 6-2. Beardsley Ruml (University of Chicago Archives)

es in scientific thinking. The memorial supported social science, Ruml later asserted, because "of the belief that the practical attack on social problems is the scientific attack broadly conceived." He and his colleagues felt "more understanding was needed than could be obtained from an appeal to tradition, expediency, or intuition." While he declared that "the end [to be pursued through scientific means] was explicitly recognized as the advancement of human welfare," Ruml seemed to overlook the paternalism implied by this outlook. Science conferred sufficient authority to make his plans for "the advancement of human welfare"—which retained existing relations of authority and power—more palatable and opposition to these programs less credible; Ruml noted that "practical significance of the results of the social sciences on public opinion must not be overlooked." "The results," he continued, "of investigations in the social sciences, where they are conducted by *obviously impartial* scientific agencies and where these results are generally accepted by scientific men, come to play a definite and wholesome part in the thinking of people generally."[16]

In contrast to the scientific ideal of open investigative communication, however, the Rockefeller philanthropy made many secret donations to social research. The *Encyclopaedia of the Social Sciences*, for example, received $600,000 from the LSRM and the Rockefeller Foundation social science division, but nowhere was this fact publicized. Ruml's attitude resulted from an understanding that the Rockefeller name be kept separate from the consideration of political issues. The attitude was not surprising in light of the foundation's history. After labor war erupted at the Rockefellers' Ludlow, Colorado, mine in the 1910s, grandstanding legislators and sensationalist journalists had reacted loudly to the family's attempt to charter their foundation, and the Rockefellers wanted no further such headaches. The efforts of a major capitalist philanthropic enterprise to confer benefits upon humanity while also existing beyond public accountability entailed many complications. Given these pressures and complexities, Ruml understandably practiced what he thought was discretion. At the same time, the oft-declared objectivity of social science served to depoliticize the memorial's political action.[17]

Ruml's new plan for an integrated social science program attacked social problems from several sides, each with a parallel in the natural sciences and engineering. The LSRM supported so-called pure research, consisting of inquiries in search of basic data on issues like crime, immigration, and employment, at major universities and at independent bureaus. Fellowships funded promising graduate students in an effort to lure talent to the project.

Accumulated knowledge could also be codified for social scientists and the general public, as it was in the *Encyclopaedia of the Social Sciences*. Principles discovered by research would subsequently be applied in practice by "social technicians" from schools of social work, public administration, and business management, and these programs also received Ruml's attention. His LSRM program thus systematically addressed investigation, dissemination, and application of social knowledge.

The SSRC, a coordinating group, performed multiple functions within the Ruml program. It promoted professional prestige, serving as an analogue to the hard-science NRC, to which only anthropology and psychology were admitted as sciences. The SSRC also provided a body of consultants to Rockefeller philanthropic managers searching for worthy grantees. As a distributor of some of the millions of dollars Ruml was spending on social science, the SSRC effectively removed the Rockefeller name from social commitments that could be perceived as political. Finally, social data could be compiled by the SSRC on a national scale and delivered to policymakers, as in the case of President Hoover's Research Committee on Social Trends.

Even as it served as a powerful component of Ruml's program, the SSRC illustrated the inconsistencies that scholars could encounter in pursuit of apolitical relations to politicized subjects like wealth and poverty or crime and punishment. The academics in the SSRC frequently sought to improve public perceptions of the investigator: image making and the desire for prestige never failed to influence the organization's agenda. But awareness of financial realities dampened self-congratulation on scholarly objectivity. The 1928–29 annual report of the SSRC declared that "the Council offers this new school of philanthropists an effective medium through which to invest their funds for the development of social science." As a result, researchers could feel pinned between the horns of the social science bull; circumstances forced them to choose between short-term social problem solving and patient accumulation of facts. The foundations' interest in results tended to encourage application over investigation, but not decisively so.[18]

When the Rockefeller Foundation reorganized in 1928, it further committed itself to Ruml's perspective even as he was leaving his post. A committee of trustees assessed the LSRM creed to see if it should be carried over in its successor institution, the Rockefeller Foundation division of social science. Although approval for the plan appeared preordained because the committee of review included Fosdick and Arthur Woods, a close associate of Herbert Hoover, dissent could be detected about the role of so-called science in shaping public affairs. One staff member wrote Fosdick that the presumed experimental attitude proclaimed by the social scientific advocates cloaked an unspoken defense of the status quo. They had, wrote Thomas B. Appelget, "tended to assume too completely the sanctity of certain fundamental institutions." The most significant of these, from the standpoint of political economy, was of course industrial capitalism, but Appelget himself could not voice this challenge. "Personally, I should be the last man in the world to doubt the usefulness of our present concepts of marriage, the family and the state," he continued. Even so, Appelget asserted that these arrangements' "present shape and character may not represent the ultimate in their development."[19]

Other authorities responded to the committee of review with guarded praise for the Ruml program. George E. Vincent, the president of the Rockefeller Foundation, raised the concern that the terminology might be misinterpreted. While "it is legitimate enough to use the analogies or metaphors which are applied in such terms as 'social clinic,' 'social laboratory,' 'social technology,' 'social engineering,'" Vincent expressed doubts about those who "too literally interpreted" the terms. Still more reviewers of the Ruml program came to similar conclusions. Henry S. Pritchett, the president of the Carnegie Foundation for the Advancement of Teaching, queried the theoretical soundness of the plan. He pointed to "certain assumptions that need to be tested, and certain fundamental questions that need to be thought out" before social science as social engineering could be accepted. Pritchett explicated the central question on the minds of social scientists across America: "Why have we not made the same progress in the social sciences during the last half century that has been made in the physical sciences?" Unlike most of the true believers, Pritchett posed a disconcerting hypothesis. He wondered if "the so-called social sciences are not sciences, in the strict sense," and if "we are seeking to apply the methods of physical science to phenomena that are not amenable to such treatment." He asserted that the "term science, in this modern sense, is stretched too far." While not condemning the Ruml program, neither did Pritchett lend it wholehearted support.[20]

These minority opinions seem not to have been reflected in the evolution of Rockefeller social science. The committee of review recommended the following, which the trustees adopted:

I. The general purposes of the program are to (a) increase the body of knowledge which in the hands of competent social technicians may be expected in time to result in substantial social control; (b) enlarge the general stock of ideas which should be in the possession of all intelligent members of civilized society; and (c) spread the appreciation of the appropriateness and value of scientific methods in the simplification and solution of modern social problems.

Such a mission was clearly political, yet the phrasing made the Rockefeller social scientists into apolitical technicians. But how was competence to be determined? Whose version of "intelligent" and "civilized" counted? The presumed objectivity of scientific method rendered such issues irrelevant; politics could be replaced by a quest for competence. Transnational, impartial, and, most important, successful, applied science got results in the control of nature. It could, these administrators felt, deliver similar success in the control of society.[21]

Edmund E. Day (1883–1951), Ruml's successor in the Rockefeller structure, loudly advocated applied social scientific knowledge. He came into the foundation from the University of Michigan business school and continued to operate on the international and interdisciplinary scale characteristic of his wunderkind predecessor Ruml. He warned his staff that "if we cannot effect anything like substantial control on the basis of scientific study of social phenomena . . . grave doubts about the possibility of overcoming" the social problems typical of industrial society must arise. Day committed Rockefeller resources to the development of methods of social investigation and control analogous to science and engineering.[22]

Day found himself in close agreement with Ruml on this central point: science had to be validated by results. Even more than Ruml, Day made his commitments plain. He told the Rockefeller Foundation trustees that if they could organize communities to "proceed analytically in dealing with the problems of the community, working with the technical expert on one side and the practical administrator on the other," the philanthropy could "create what is essentially a new social order." In contrast to some conceptions of "pure" science that stress knowledge for its own sake, Day repeatedly claimed that "the validation of the findings of social science must be through effective social control." This control would inevitably enhance the lives of the administered. When pressed on the point at a trustee meeting, he countered, "Perhaps I have rather rationally assumed that the development of a more scientific approach to the social problems would yield a substantial humanitarian product." He continued, spelling out what he thought was obvious: "Underlying the whole conception, of course, is a very definite humanitarian purpose. I can't see how a substantial advance made along this line

Fig. 6-3. Edmund E. Day (Michigan Historical Collections, University of Michigan, Ann Arbor)

would fail to minister to the well being of mankind throughout the world." Such ambitious possibilities remained unattained, despite a "New Social Science Program" in 1933 that stressed "a frank shift of emphasis to concrete fields of application." In apparent frustration, Day left the Rockefeller Foundation in 1938 to become the president of Cornell.[23]

. . . .

Day and Ruml oriented the Rockefeller research and enrichment programs toward rational reformation of society and politics, and social science practitioners soon learned the benefits of adopting—or appearing to adopt—the prevailing viewpoint. Charles Merriam and Wesley Mitchell led a vanguard of institution builders who saw social science as both epistemologically objective and politically useful. Both reinforced growing networks of academic-governmental-philanthropic cooperation and developed persuasive and well-respected justifications for social investigation along engineering lines. They also were among the few academics who attempted to put theory into political practice as they headed the National Resources Planning Board under Franklin Roosevelt. Even before the New Deal, however, Merriam and Mitchell actively championed the social investigator as social engineer.

How, asked Charles Merriam (1874–1953), "may political science make best use of all that the other sciences are contributing to modern thought and practice?" In answer to his own question, Merriam served the cause of applied social science within several institutions: his own University of Chicago; the SSRC, which he helped to found; and the Recent Social Trends venture. He also modernized the field of public administration, trained a generation of leaders in political science (among them Harold Lasswell and V. O. Key), organized important conferences, and helped to introduce psychological conceptions of political behavior into his discipline. A close associate of Ruml, Merriam later sat on the board of the Spelman Fund, a more narrowly focused successor to the LSRM. In all of these settings, Merriam attempted—with varying degrees of success—to balance public appreciation of and education in the ways of science with centralization of authority among expert administrators.[24]

In addition to providing institutional leadership for advocates of applied social science, Merriam developed intellectual justification for political scientism. In this view, outlined most influentially in *New Aspects of Politics* (1925), empirically derived knowledge foreshadowed scientific social control. The model of engineering practice thus was implicit in all of Merriam's ventures. In a widely noted address to the APSA, he confidently predicted

Fig. 6-4. Charles E. Merriam (University of Chicago Archives)

"that we may definitely and measurably advance the comprehensiveness and accuracy of our observation of political phenomena," and that the mechanisms "of social and political control may be found to be much more susceptible to human adaptation and reorganization than they now are." Merriam's friend Morris Cooke reinforced this outlook, and the two men worked to link sessions of their professional organizations, assuming the political scientists and "Efficiency Engineers," as Merriam called them, shared common ground. Like Day, Merriam announced that his ideology validated inquiry by results: "Constructive intelligence in social affairs," he wrote in 1926, "is the final end of social science."[25]

For many years Merriam wanted a national conference of businessmen, academics, and other leaders who could conduct ongoing discussions about making politics more scientific. Such a project was quintessential Merriam: it appealed to foundations for substantial funding, it educated citizens in the ways of scientific investigation, and it had a directorate as well-connected as the Brooklyn Bridge. The fundamental premises of the conference also bore Merriam's mark: because the United States failed to make good on "the promise of efficient democracy," new solutions were required. Engineering and managerialism supplied the new blood the ailing body politic demanded. Because America had "combined intelligence, men, and materials to form the most efficient business organizations the world has known," the question became one of doing the same with government. Merriam posited that "the highway of progress is paved with inventions" and that "invention is the reward of research." But, he concluded, "Can research do for government what it has done and is now doing for science and industry?" Significantly, the project, which never came off, would have placed a consulting psychologist and a consulting statistician on each citizen roundtable.[26]

Merriam's version of scientific social control, however, did not necessarily imply elitism; science had to reshape education in order that citizens could apply its lessons to their politics. With advanced technology making shorter workdays possible, citizens would have more time to be educated about and active in government. Responsibility for this training in scientific understanding fell on the social scientist, according to Merriam. He called it "essential" to devise "a means by which the public may be kept in touch with what is going on in the scientific world." Although Edwin Slosson's Science Service provided popularized accounts of developments in natural science to newspapers and magazines, a similar enterprise "is even more necessary in the case of social science." Given appropriate education and renewed commitment to social scientific investigation and application, the future, imposing but full of possibility, beckoned. "It is difficult to see," Merriam wrote, "how the new world can continue, endowed with scientific mastery over the forces of nature, yet with a government ruling by the methods of pre-scientific time." Through the interwar years Merriam continued to equate scientific government with modernized democracy, although he faced stiffer opposition all the time.[27]

Perhaps even more than Merriam, Wesley C. Mitchell invested substantial faith in scientific inquiry and application, attempting to have matters two ways. He advocated both objective, impartial investigation and salvation from social problems through application of scientific precepts. His institutional affiliations proliferated: Mitchell helped found the New School for Social Research, the NBER, and the SSRC and also headed the Recent Social Trends project. He developed working relations with the philanthropic managers, with Herbert Hoover, and with Franklin D. Roosevelt, thereby gaining an access to power not usually available to academic researchers. Throughout his career, an analogy drawn with natural science informed his endeavors. Institutes for research in the social sciences, he asserted, "promise to become as authentic an agency for promoting human knowledge and human welfare as the laboratories from which physical science emerged to make over the world." Similarly, Recent Social Trends, "a venture by the President to apply the procedure of social technology to social problems," was "a hopeful portent to all who believe that man's best chance of bettering his lot lies in using his brain."[28]

In the aftermath of World War I, Mitchell had linked the possibility of human progress to the success of social science in imitating applied science. A letter to the *New York Evening Post* in 1920 had argued that "social technology may achieve a cumulative growth like that of mechanical technology," while in an address to a church group in the same year Mitchell posited that "so long as the social sciences remain in this [backward] state, they cannot serve as a guide to action in political matters as natural sciences serve as a guide in technical matters." Once again, knowledge could only be validated by its capacity to influence human existence. Mitchell's early training under John Dewey and Thorstein Veblen contributed to his thought an ongoing commitment to usefulness; it seems fitting that Mitchell the superstar theorist relaxed by making wooden furniture. "We desire knowledge," he wrote in 1924, "mainly as an instrument of control." That mastery descended directly from engineering: "Man's effort to understand and control nature be-

Fig. 6-5. Wesley Clair Mitchell (NBER)

comes an effort to understand and control himself and his society." Scientific investigation into human institutions could, he believed, yield tangible social benefits.[29]

Like Merriam, Mitchell cultivated friendships with enough engineers and managerialists to know their aspirations. He addressed a meeting now inconceivable: a joint session of the AEA with the ASME in 1922. Mitchell's understanding of his own profession, in fact, drew heavily on the heritage of scientific invention. In one essay he began by noting the friendship between Adam Smith and James Watt at Glasgow College. From that point, history was forever changed: "In retrospect it is clear that the economist and the engineer were working to a common end. Both these Scotchmen had a plan for increasing the efficiency of production," Mitchell wrote, apparently reading *The Theory of Moral Sentiments* somewhat idiosyncratically. Since that time, he concluded, "the trend of economic research has brought it nearer to engineering." Even so, Mitchell's continued proximity to engineering reform and managerial innovation did not confirm him as an uncomplicated social engineer.[30]

Like many other social scientists, Mitchell struggled both to solve current problems and to compile data over time, for only the long-term accretion of knowledge would certify economics as a true science. The lure of governmental involvement and professional leadership outweighed, it seemed, the quiet and unspectacular detail work that detachment would have demanded. Nevertheless, Mitchell pleaded with foundation officials to allow him and his NBER adequate time to validate their existence without needing to produce instant results. "Of these two tests of a progressive science," he wrote Raymond Fosdick in 1927, "the test of unfolding ever new problems is more important than the practical application of results: for a narrowly utilitarian view of science defeats itself." Mitchell's pursuit of increased technical sophistication within economics, especially quantification, reinforced this desire for economics to become more authoritative than useful.[31]

Within academic sociology, no organizational and theoretical leader of the stature of Merriam or Mitchell appeared. Two men, Howard Odum and William F. Ogburn, do deserve brief mention because of their role in the application of social scientific knowledge during the 1920s. Both served in the Recent Social Trends leadership and achieved professional importance because of their understanding of the possibilities of social science.

William F. Ogburn (1886–1959), known throughout American social science for his theory of cultural lag—a notion that drew heavily upon the work of the much less orthodox Thorstein Veblen—also devoted substantial work

Fig. 6-6. William F. Ogburn (University of Chicago Archives)

to the definition of social science. For him, application was to be divorced from investigation. Despite working on a Recent Social Trends volume widely viewed as a blueprint for action, Ogburn directed a steady stream of caution to douse colleagues he thought to be inflamed by the prospect of usefulness. The most visible of these warnings, "The Folk-Ways of a Scientific Sociology," outlined in unambiguous terms the conditions for scientific authenticity. "Sociology as a science," the article declared, "is not interested in making the world a better place in which to live . . . or in guiding the ship of state. Science is interested directly in one thing only, to wit, discovering new knowledge." Application of findings, however tempting, would be left to statesmen and social workers in order that scientific purity could be maintained. Although "it will be necessary to crush out emotion and to discipline the mind so strongly that the fanciful pleasures of intellectuality will have to be eschewed in the verification process," such an astringent world would provide the investigator with a pure, unsullied refuge: his laboratory. Suspicions that Ogburn's psyche was perhaps too tightly wound appear reasonable: his behavior as director of research for Recent Social Trends revealed a sometimes paralyzing intensity.[32]

In contrast to Ogburn, Mitchell, and Merriam, Howard Odum (1884–1954) never seemed to break into the first rank of his profession. A somewhat sloppy thinker and writer who rarely revised his material, Odum walked a difficult line in his position as director of the Institute for Research in Social Science at the University of North Carolina. He sought both to maintain the appearance of racial progressivism and to placate those who pressured the university to refrain from upsetting existing relations. His leadership of the regional planning movement in the 1930s (where he teamed with Lewis Mumford), his membership in a number of organizations devoted to racial harmony, and his work as a sociologist and editor of *Social Forces* illustrate how scattered his efforts were. As a result, his advocacy of applied social science contained all the right words, but seldom did he test his biases or elucidate his ideas with adequate complexity. These declarations are valuable for our purposes largely *because* they were not original: Odum's use of the social engineering paradigm illustrates how accessible the notion had become to reformers.

In his introduction to *American Masters of Social Science*, Odum wrote that "democracy, institutions, learning, social process, depend upon the capacity of the race to develop and utilize the social sciences as they have those in the physical realm." Without evidence, he continued: "On this point engineers and physical scientists, no less than social scientists, are of one mind." Such sweeping generalities characterize much of his writing. In 1925 Odum had proclaimed that "the need for scientific method and habit is universally applicable to all of the social problems which we have listed, as well as to the scores of others which each citizen must face." Just what the scientific method and habit were was never explained. Odum admitted that social science differed from natural science without saying how and foresaw that social scientists needed new techniques to differentiate themselves, but he gave no specifics.[33]

Fig. 6-7. Howard W. Odum (Southern Historical Collection, University of North Carolina, Chapel Hill)

Odum also bore at least partial responsibility for Hornell Hart's *The Techniques of Social Progress* (1931), a title in the ongoing series Odum supervised at Holt. From the outset, Hart's textbook took social engineering as a given and specified in detail how it worked:

What should be done when a community confronts cultural tensions calling for progressive alterations in social structure?

1) Investigate the problem

2) Bring creative minds into contact with the problem

3) Organize into effective power the diffused and inarticulate innovative energies

4) Enlist the support of individuals powerful in the institutions to be altered

5) Rouse apathetic people to interest by dramatizing the problem

6) Link up auxiliary emotional complexes with the cause to be promoted

7) Avoid needless rousing of opposition

8) Find common denominators of innovative interest

9) Demonstrate publicly and if possible dramatically the feasibility and usefulness of the innovation

10) Spread everywhere information as to the best methods and equipment used anywhere.[34]

Despite social engineers' fascination with the empirical, Hart gave no examples of how this self-contradictory (see numbers 4 and 7, or 5 and 7) and cumbersome program had been accomplished by citizens in the flesh. As in Taylorism, the elegance far outran any realistic possibilities of implementation. The authoritarian side of the ideology, bereft of apologies, was justified in its pragmatic adequacy. Social engineering solved problems, so the main task for the present became the "transfer [of] the power of choice to wiser and more public-spirited leaders." Note also that Hart listed emotional energy, ordinary citizens, and dramatic portrayal in his program in contrast to other rational reformers' antipathies toward these vestiges of "politics." Similar declarations appeared frequently in *Social Forces*.

Once he detected the prevailing ideology at the LSRM and won significant funds for his social science research institute, Odum became a nearly constant correspondent with the social work scholar Sydnor Walker and with Edmund Day in particular; similar barrages landed on the publishers of Odum's many and often forgettable books. Seeing himself as an institution builder and scholarly mover in the Mitchell and Merriam mold, Odum lacked the firepower to pull off such an impersonation. That said, plenty of investigators took *Social Forces* and the UNC research institute seriously, and Odum's influence on racial liberalism was, if not terribly vigorous, at least an effort to stir things up. In the 1930s he maintained his facile faith in social science long after many had raised serious doubts and pursued projects aimed more at upbuilding his prestige than at developing a coherent sociology.

....

In the postwar decade, a few influential academics and administrators founded a network of institutions devoted to rationalizing social progress. They almost always assumed the theoretical underpinnings for such a pursuit to be self-evident, though frequent references to Karl Pearson's *Grammar of Science* provided a patina of scholarly justification in some cases, more substantial foundations in others. The language of engineering, still coexisting with medical, household, and other metaphors, remained nearly ubiquitous in these rarefied circles. Most reformers remained reluctant to question the implications of their mechanism and positivism for social, political, and economic life. By 1925 the institutions had started to apply social science and thus act on their theoretical commitments. These ventures, especially, illuminate some of the frustrations and miscalculations that the erstwhile social engineers encountered.

7

SOCIAL ENGINEERING PROJECTS: THE 1920S

The LSRM under Beardsley Ruml and its successor, the Rockefeller social science division headed by Edmund E. Day, supported many efforts to collect, analyze, and disseminate empirical social facts in order that new knowledge could be applied to American social problems. Two of these ventures—*Recent Social Trends* and the *Encyclopaedia of the Social Sciences*—stand as monuments to a golden age of social science professionalization. A third project—annual meetings of the SSRC—reveals some of the difficulties that scientism evoked among academics. Many competing ideologies and outlooks complicated all of these ventures, and I am reading the episodes only for the influence of the engineering model; to suggest that the *Encyclopaedia*, *Recent Social Trends*, and the Hanover conferences were solely exercises in social engineering would be a mistake. In these projects, however, scientistic yearnings for neo-Comtean *social control*—the phrase that succeeded *efficiency* and anticipated *planning*—clearly animated much thought and action.

THE HANOVER CONFERENCES

After Ruml assumed his duties at the LSRM, one of his earliest innovations involved six annual summer retreats at Hanover, New Hampshire, where he had attended college at Dartmouth. Guests and members of the SSRC were invited to spend, in the early years of the Hanover conferences, the weeks of August in residence, lodged in a campus fraternity house. A typical day consisted of mornings devoted to meetings of the various SSRC committees (especially the Program and Policy committee, which in some years got a week to itself), unscheduled afternoon recreation, and evening sessions that brought together all participants to hear and discuss a guest speaker's views on some issue of moment. This experience was conceived to be more than a session of informal contact and cross-fertilization, as Ruml pointed out at the initial meeting in 1925: "The Conference has been organized not so much for the purpose of securing an interchange of ideas as for the purpose of creating a vision from which new ideas may emerge." Modes of reasoning borrowed from natural science and engineering shaped many participants' understanding of the future's best path to social progress.[1]

Because they broadened institutional networks, the conferences performed an important function; the SSRC was devoted to the encouragement of interdisciplinary cooperation, and the inclusive invitation lists helped both to spread social engineering ideas and to discourage intellectual isolation. Ruml and his staff arranged for many social investigators who had demonstrated impulses toward scientific social control to be invited to Hanover, and the conferences served to acquaint kindred spirits in a congenial setting. Social work and industrial management, in particular, sent representatives like Mary van Kleeck, Harlow Person, and Henry Dennison who compared notes with academic investigators. Many people came to Hanover in the summer: Mitchell and Merriam; Odum and Ogburn; the psychological testers E. L. Thorndike and Elton Mayo; and economists including Edwin Gay of Harvard, Walton Hamilton and Edwin Nourse of Brookings, Adolf Berle, and Alvin Johnson of the New School. Roscoe Pound and Felix Frankfurter attended, as did Charles Beard and then–SSRC staff member Robert Lynd. The heyday of the affair appears to have been 1926, when about one hundred social investigators and related persons attended the conference, which ran, in two sessions, from August 9 until September 3.[2]

In addition to social investigators, the foundation administrators themselves comprised the other important constituency at Hanover. From the early days of the SSRC, philanthropic officials including Shelby Harrison and Mary van Kleeck of Russell Sage and Day from Rockefeller had held committee and other administrative posts, so their inclusion at the conferences made sense. Other foundation officials such as Carnegie Corporation president Frederick Keppel attended as well, as did a cadre of Rockefeller social science administrators that included Ruml, Lawrence Frank, and Sydnor Walker. Some investigators resented them; Robert Lynd noted that "a research worker spoke of the chilling effect at Hanover of our having foundation people and administrators in the crowd." More common, apparently, were grant seekers who took advantage of golf outings and wilderness hikes to impress upon captive foundation officials the merits of their projects. Merriam kept whimsical notes on his visits, and in one undated entry he observed that "the Foundations assemble, for the first time—in a zone of quiet, in an atmosphere of Mystery—the mouths of Council members water, but they are discreet, observing the promise of 'safe conduct' for the Philanthropists." Researchers and foundation administrators undoubtedly alternately resented and appreciated the informality and proximity of the gatherings.[3]

In keeping with standard operating procedure for Ruml and the Rockefeller social science ventures in general, the men and women of the SSRC developed their conceptions of social engineering in both isolation and secrecy. SSRC president Arnold Bennett Hall told the opening meeting of the 1926 conference the rules of the game: "To get some of the men we have here as well as to facilitate generally our proceedings, we have had to insist that it is to be absolutely confidential." In concrete terms, Hall declared, "nothing is to be given out to the news and no public statements made." Even as the administrators and investigators sought to transcend politics with science, they tacitly acknowledged that the meetings constituted political, and very possibly controversial, activity, and took no chances as to the wrong impressions being drawn.[4]

The growing coziness of social investigators, universities, and philanthropic foundations generated significant political implications; it took an outsider to call that world to accountability. In the August issue of *Harper's* that appeared just before the 1928 meeting, Harold Laski of the London School of Economics skewered social science and its patrons. Commenting on the whirlwind of SSRC efforts, and therefore Rockefeller efforts, Laski pointed to "its grants-in-aid to the established, and its fellowships to the immature"; to "interim reports, special reports, confidential reports, final reports"; and to "experimental centers, statistical centers, analytical centers" as evidence of a debilitating systemization of research. Cooperation, Laski argued, "has never adequately replaced the vision and the insight of the individual thinker." He also attacked the "bright young, or pompous old, executive of some foundation, to whom the very meaning of research is, in any effectively creative sense, entirely unknown."[5]

The price of such men dictating the agenda of research, Laski feared, was one "we are ill able to afford." Political thought would have to give way to superficially apolitical social technique. "Because scientific 'impartiality' is important—for the donors must not be accused of subsidizing a particular point of view—the emphasis of research moves away from values and ends to materials and methods," he wrote. Even though "Plato, Hobbes, Rousseau, Hegel, Bentham, do not seem to have been natural members of committees," that mode of organization had captured the moment. More important, self-criticism had diminished in the general aura of grantsmanship and self-congratulation: "If somewhere a faint doubt obtrudes, a reference to the technic of the natural sciences and the immense results secured there is usually sufficient to stifle skepticism."[6]

Before the discussion of the importation of natural science methods into the world of social investigation could resume in 1928, the faithful had to enforce orthodoxy by condemning the heretic. Harold Moulton of the Brookings Institution took the floor and dismissed Laski as being "dead wrong," insisting that funds "have got to come from somewhere." When "we just raise our hands," and decry the influence of the foundations, he contended, "we are just kicking against the bricks." Self-satisfaction continued to rule, Laski's basic challenges having gone unmet.[7]

Laski saw the transcript of the meeting at which he was superficially bashed, remained unconvinced, and told Felix Frankfurter as much. The jurist agreed in a letter to Fosdick:

> There is room for the organized achievement of statistical and informational data and for the encouragement of discussion by people working in kindred fields. . . . But those aren't the matters that are of real moment. The deeper questions are what will encourage originality of thought; what are the unconscious and even, if you please, the unintended influences of subordination and subservience in those who have the guidance of great institutions; what makes for the essential free-spiritedness on the part of academicians and what makes against it; what are the forces in academic life that are over-emphasized and what are undervalued; what makes for long-term inquiry as against immediate "practical" results; what will make for the right kind of people going into academic life and what will keep the wrong ones out? If these and other such questions were candidly and vigorously grappled with at Hanover, the minutes which you were good enough to send at all events do not disclose it.

Frankfurter's insight was exceptional, for such dissatisfaction with both means and ends seldom appeared in this period, especially among American social scientists.[8]

Rather than engaging the larger issues involved in any conception of social control, Hanover conferees attempted to define how they could make social inquiry and application scientific and so enjoy success comparable to that of natural science and engineering. Mitchell mentioned to an evening session in 1926 that "the subject of scientific method is one that we seem continually to be discussing," while Merriam put the state of things in characteristically offbeat terms. He told the 1927 conferees about the SSRC's interest in "relations between natural science and the social sciences, in what are sometimes called the natural and unnatural sciences." This version of science, however, was not defined analogously to astronomy, where the subjects of research were to be observed from afar; rather, "we have looked toward social intervention and in a broader sense toward social intelligence as a factor in social affairs." To be valuable, knowledge had to facilitate control.[9]

As a result of the ongoing debate, the SSRC named a Committee on Scientific Method and instructed it to publish a discussion of the current state of the field. In a planning document of 1927, the committee explicitly affirmed its origins in natural science. Such a definition made the transition from social science to social engineering possible, indeed inevitable: "In every scientific field, the facts or principles established by scientific enquiry may *legitimately* be applied to the service of human purposes or shown to have a bearing on human problems." For example, "If astronomy is applied to the art of navigation, why should not economics be applied to the art of saving and spending, and political science to the art of governing?" The committee's concerns reappear throughout the Hanover transcripts. Clearly the matter of means and ends raised issues that had implications across the entire SSRC membership.[10]

Part of the control conferred by the application of science would come through the scientific education of the masses. At Hanover, a 1927 conferee declared that "when maladjustments in human behavior occur, we should look for real and controllable causes of them and get the common man into the same habit." In 1926 the psychologist Floyd Allport had expressed similar reasoning: "What we need to do is to stimulate not only research workers in universities to think and work in scientific terms, but also the people at large, perhaps not to work scientifically but to think more scientifically about social problems." Words like *maladjustment* imply some norm, defined by unmentioned processes, so some participants wondered how much ethical content could appropriately leaven applied social science.[11]

Hanover social scientists made frequent declarations of their social obligation as a matter of course. Allport admitted that his assertion "raises a question of whether there will be an ethical note at all in our orientation," while Merriam took another step. While the SSRC "must protect the highest standards of most precise scientific attainment," he noted in an undated Hanover document, "science cannot escape social responsibility, by silence or refuge in superiority." Such imputations made Edmund Day wary, and he hedged his position; asserting that ethical awareness was a bad thing would sound indecorous. On the other hand, social inquiry en route to social engineering needed the credibility of science, not merely a sense of uplift. "The findings of economics might be more seriously regarded," he posited, if they came from "men who hewed pretty closely to the line of scientific work, didn't give advance [advice?] and didn't mix their findings of fact with a lot of ethical considerations." Later in the discussion, he summed up by asserting that "social science suffers immeasurably from the fact that it is a composite of scientific, philosophical, and ethical material."[12]

As Day expressed concern for the mingling of fact and value within the broad rubric of science, the conferees often celebrated science not as a body of material but as a method and as a spirit, with limitless potential for human improvement. In the notes cited above, Merriam pointed to "violence and propaganda" as possibilities for the future. Science, on the other hand, presented humanity with an alternative vision: "Modern science—modern research—modern intelligence—may pour oil on the troubled waters. Or may avert the ruder shocks of change. Or may supply such guidance in crises— that the lot of mankind may be happier than ever before, the gains of science may be realized for the enrichment of human life, creating new systems and new values, more vivid and satisfying than have yet appeared." This hopeful perspective did not, however, convince the SSRC members who wanted to do "real" science.[13]

Unreconstructed by such buoyancy, Merriam's University of Chicago colleague William F. Ogburn controverted both social ethics and social engineering. As "The Folk-Ways of a Scientific Sociology" illustrated, his conception of the scientist included a remarkable capacity for distancing oneself from the immediate social context. "Science is the accumulation of knowledge, and knowledge thought of in a very exact sense of the word," he told the 1928 meeting. Declaring that social science could not aspire to "social engineering of a very high measure until you get the relationships measured," Ogburn called for sharper focus on data-gathering for its own sake: "If the pressure is put upon solving the practical problem, upon the ethical point of view, upon the choice, or upon doing something that is worth while, you obscure very greatly and hinder very greatly the process of getting this information." Science, for Ogburn, thus *was* closer to astronomy than to engineering, putting him in an adversarial position with regard to certain reformist forces. At the same time, his view lent valuable credence to Day's methodological orientation.[14]

Another Chicago social scientist, Frank Knight, also scrutinized the prevailing social scientific outlook. He unrelentingly questioned what science could and could not become while making many of the ethical implications of the natural science and engineering analogies plain and thereby troublesome. His presentation at Hanover in 1928 did not originate on the spot, for as early as 1924 he had been raising similar concerns. By investigating the epistemology and even the metaphysics behind the advocacy of social science, he raised perplexing issues.

Knight's program, a modified Jamesian pragmatism, favored common sense over ritualized social science, appreciated meaning found in process rather than results, and respected individual choice more than social control advocates did. In his essay of 1924, "The Limitations of Scientific Method in Economics," Knight forcefully opposed the increasing scientism apparent in organizations such as the SSRC. Human existence, he asserted, "is at bottom an exploration in the field of values" rather than an attempt to fulfill those values materially; "we strive to 'know ourselves,' to find out our real wants, more than to get what we want." Knight opposed what he called appreciation, an ongoing process, to creation, which consists of shaping new values and new means for the realization of those values. Social technicians, though, tend "more and more to subordinate the desire for understanding as such to a desire for control."[15]

Making everyday life "scientific" thus unnecessarily complicates matters, Knight argued. In their quest for objectivity, advocates of scientific reasoning ignored underlying value questions: scientific methods do not tell people what they want, only how to get it. Instead, the concept of usefulness, assumed to be an apolitical valuative criterion, is in fact a metaphysical concept smuggled into the discussion. To make everyday decisions, humans rely on common sense which is not codified or ritualized: "It is all in the field of art, and not of science, of suggestion and interpretation, and not accurate, definite, objective statement, a sphere in which common sense works and logic falls down." Careful not to be confused with a Luddite, Knight proposed "not that prediction and control are impossible in the field of human phenomena, but that the formal methods of science are of very limited application." He thus resisted the application of science to society on both functional and ethical grounds; even utilitarian concerns favored a more humanistic mix of art and science.[16]

In this analysis, the epistemological and moral premises of scientific method go bankrupt in relation to human experience. Because "meaning, being a subjective phenomenon, must be suggested rather than stated," social science faced a crucial methodological problem: no "language, except possibly the arbitrary symbols of mathematics and symbolic logic, is entirely literal." The critique continued onto other grounds. What, really, is "social control"? Knight asked. "In practice," he answered, "this can only mean either something on the order of self-control . . . or else the control of one part of society by another part." Given that the ideal of American society is ostensibly freedom, natural science once again provided a poor analogy: "Physical objects are not at the same time trying to understand and use the investigator!"[17]

At Hanover several years later, Knight amplified this argument. In an evening session lasting until after midnight, he defended his position against many of the leading advocates of applied social science, including Day. The level of discomfort becomes readily apparent in the stenographic typescript of the meeting: opponents frequently interrupted his presentation to quibble, and much of the later discussion revolved around issues of semantics rather than content. Like Laski, whose attack came out just before this conference, Knight struck home on matters of professional importance largely ignored by the SSRC.

His critique at Hanover once again centered on the moral and epistemological issues involved in social engineering. The perspective remained the same: "Social science has been getting off on the wrong foot by attempting to ape the natural sciences." What he called a "natural science hypnotism" led social investigators to distrust their endeavors unless they were "using the concepts and formulation of the natural sciences," even though simple reflection would show that the two realms differed fundamentally. The orientation imported from the natural sciences led the investigator to approach his subject inappropriately. In natural science, Knight maintained, "we are thinking about an external world." Because in the social sciences "we are largely thinking about ourselves," what results is "an ultimate contradiction, an ultimate mechanical contradiction, that we somehow transcend." That transcendence made the social scientist a "*deus ex machina* looking on from some other world." The inability to know one's medium as through a microscope challenged the heart of the natural science analogy.[18]

Social science, as understood by many practitioners, involved not only comprehension but application, and Knight attacked *control* as "a very ambiguous word," one "much misused." Social science as social control confronted a particular difficulty in America. According to Knight, "Government by consent [of the governed] makes a categorical difference between the type of control exercised by government and the type of control exercised by man . . . over the natural world." As a result, social science needed to look elsewhere for its methodological models. Instead, Knight suggested the relationship between student and teacher because it dealt with reciprocity, speculation, and interpretation.[19]

Finally, in terms similar to Laski's, Knight explicitly worried about the role of the foundations in relation to academic scholarship. He exaggerated the threat somewhat when he warned that the centralization of social research "is going to have a tendency, if it isn't very carefully watched, to lead into the regimentation and mechanization and materialization and organization of life." He also worried that a focus on inappropriately applied science caused particular problems in the America H. L. Mencken so distrusted, which was "notoriously . . . not a favorable environment for the growth of art and the finer social sentiments and graces." Knight thus occupied a lonely position holding to enlightened humanism. He sought to escape from the past represented by William Jennings Bryan at the Scopes trial while still not uncritically accepting the viewpoint that science could solve everything. That position, unfortunately, turned out to be only marginally solid middle ground, and the extremes of the moment—scientism and fundamentalism—made discussions of social values difficult. Knight could decry the dangers of scientism, but he could not root a system of moral values in a "progressive" alternative to a rapidly ossifying Judeo-Christian intellectual tradition.[20]

Not surprisingly, Edmund Day diametrically opposed this outlook; as Merriam noted, "Knight and Day seem far apart, without much twilight." Day's Hanover speech, "Trends of Social Science," reveals the divergence in stark terms. With the weighty prestige of the Rockefeller philanthropy backing him, Day attempted to set the agenda in decisive terms. One suspects he was less shaping conceptions of science than articulating conventional wisdom, but in either case he served as a powerful exponent of scientism.[21]

Day encouraged social scientists to pursue both methodological and professional goals. With regard to the former, he initially insisted that "the nature of social philosophy, or of social ethics, does not here concern us." Instead, he defined "social science as the systematic purposeful study of social structure and social process." That study had to be validated by results—findings must facilitate "the more intelligent understanding and the more effective manipulation of social force." This style of testing inquiry by results posed certain problems, for Day asserted both that "there must be a purpose behind the inquiry" and that "even when there is a measure of self-interest in the situation, we may expect our social scientists none the less to be objective." At the same time, value questions unnerved him, so he called for "more tough-mindedness in social science": too many practitioners were "preachers who have lost their theology." Thus the proper social scientist, both interested and disinterested, needed self-control, or schizophrenia, in regard to goals and values.[22]

In an appeal to the group's sense of professional importance, Day portrayed the SSRC of the future in dynamic and vital terms. For that future to emerge, amateurs had to be excluded from the consideration of society's direction. "We need to convey to the people on the outside the notion that social science, competently conducted, is highly technical. It always irritates me to have the impression conveyed that anybody can be an economist if he just talks about economic subjects," Day told the gathering. If social scientists did not grasp the possibilities of social control, terrible things would happen: "Social science as an effective instrument is going to . . . fall into the hands of those who are commonly regarded as statesmen or politicians or administrators or business men." Day waved a red flag in front of the SSRC bulls. No threat of political radicalism would worry him; instead, businessmen and statesmen populated his personal hell. The hope that social science could transcend mere politics could be no more bluntly stated.[23]

The social scientists themselves did not follow the letter of the foundation law, as Ogburn's stance would suggest. In an SSRC planning meeting in 1929, A. B. Hall argued for a higher public profile on the part of American social scientists in order that scientific advice on social issues would be heeded within conventional styles of political discussion. "We talk about social engineering," he said, "but obviously that is wholly an academic question in a democracy unless there is public appreciation of what social engineering can mean in the formulation and execution of policy." This program of public relations, developed in nonspecific terms at a meeting with Day in attendance, came to little within the larger cosmos of the SSRC's activities. By the late 1920s, academics detected some of the discontinuities between democratic government and privately supported social efficiency.[24]

Laski's and Knight's objections to the Ruml-Day program failed to inspire the SSRC to reassess philanthropy's relations to applied science and to democratic government. They do, however, stand as historical forebears to the wave of antiplanning sentiment that followed the rise of totalitarian scientific planning in the Europe of the 1930s. By 1939, Laski and Knight stood much closer to the mainstream of social opinion, and the planners fought a rearguard action while philanthropies redefined their mission and programs. After 1940, advances in atomic science in particular would distract public attention from engineering achievements, and the analogy had to be reworked between the "natural and unnatural sciences."

THE *ENCYCLOPAEDIA* OF THE SOCIAL SCIENCES

During the interwar period, the *Encyclopaedia of the Social Sciences* took shape at the nexus of private philanthropy, scientific social reform, and the quest for professional prestige. Associate editor Alvin Johnson (1874–1971), the intellectual architect of the project, adhered to a mainstream liberalism with some scientistic overtones. Although he had worked closely with the *New Republic* editors and admired the work of Thorstein Veblen, Johnson maintained a neoclassical economic perspective, eschewing the institutional model employed by Mitchell and others closer to the center of social scientism. Despite these allegiances, his *Encyclopaedia* demonstrates, once again, the allure of engineering as a model for reform. At the same time, the set accomplished many other goals as well; scientism comprised but one facet of a complicated undertaking.

Edwin R. A. Seligman, a German-trained economist and perhaps the most eminent of American social scientists, led the project, backed by a board of directors that included such prominent advocates of applied social science as Wesley Mitchell, Mary van Kleeck, and John Dewey. Johnson, by this time dean of the New School for Social Research, agreed to accept the post of associate editor and in turn assembled editorial assistants, formulated subjects, selected authors, and set editorial standards. A letter that Johnson wrote later in his life outlined many of the tasks he took upon himself:

> I was just Associate Editor; but every article in the whole work was designed by me; every article, in whatever language except the Slavic was read by me, assigned by me for translating or editing, read by me in the final version.
>
> And not one single contributor ever yelled that his copy had been badly treated.

Fig. 7-1. Alvin S. Johnson (Yale University Archives)

Lalor and Palgrave and Bliss, the Britannica and the International had all been criticized as organs of cliques. I passed over our Joint Committees, our Advisory Editors. I made lists of topics and sent them to every scholar who counted in a special branch. "Add topics that I might have missed. Strike out topics that are dead. But for every live topic suggest the most competent contributor."

Thus I consulted 600 scholars. Only one refused to cooperate. He was a Socialist and thought he ought to be paid for his opinion.

I have had more than a thousand letters from users of the Encyclopaedia. Not a single one has charged the work with bias.

After so successfully devoting seven years of his life to the mechanics of one encyclopedia, Johnson spent much of the 1950s attempting to get a revised version under way.[25]

The urbane, wealthy Seligman, meanwhile, directed most of his attention to business matters. He persuaded the LSRM to underwrite some of the production costs, although Beardsley Ruml kept his philanthropy in a background role by asking Seligman not to publicize the grant. Macmillan agreed to publish the set of fifteen volumes, an expansion on the originally projected ten, one at a time. In a most irregular manner, the firm and Seligman signed a standard contract, except that for each occurrence of the word *author*, Seligman substituted *editor-in-chief*. Had the contract remained in force, Seligman would have received 10 percent royalties for the first 1,500 sets sold and 15 percent on all sets thereafter. Considering that he never directly supervised the editorial work, Seligman stood to benefit substantially from a project made possible by foundation support; the *Encyclopaedia* remained in print into the 1960s and eventually sold over 18,000 sets.[26]

Johnson and Seligman astutely aimed their venture at several constituencies. They initially had to overcome the objection that the fields represented in the *Encyclopaedia* were advancing too quickly to be codified in anything as suggestive of finality as an encyclopedia, but Johnson answered with evidence from professional success: rapidly expanding graduate education was demanding of students broad knowledge and bibliographic control they often could not muster. Seligman explicated three further purposes in the preface. The *Encyclopaedia* was to provide "a synopsis of the progress" of the social science fields, "an assemblage or repository of facts . . . [for] all those who are keeping abreast of recent investigation and accomplishment," and "a center of authoritative knowledge for the creation of a sounder and more informed public opinion." These objectives, all met to a large degree, embodied the audacious ambition of the project.[27]

Although they saw contradictions between reform and scholarship that later scholars would solve with abstraction and specialized language, the *Encyclopaedia* organizers consciously sought to influence public affairs with scientific expertise. Felix Frankfurter predicted in 1926 that the volume would "demonstrate that social problems can be brought under the discipline of a scientific temper as rigorous and uncompromising as that which governs the so-called natural sciences." Four years later Johnson wrote Frederick Keppel that it had become "increasingly difficult for the intelligent layman to find his way about in the scientific literature." He also complained that citizens were unable to "apply the conclusions of scholarship to practical situations." The work for which Johnson sought Carnegie support would accordingly act "as a bridge from scholarship to public opinion." A memorandum on the proposed venture (most likely from Seligman) detailed the process whereby the underwriters would contribute to the development of an educated citizenry: "It ought to be a standard work of reference in every public library and in every important newspaper office, so that the fundamental ideas would gradually percolate down to the wider public. The consequence is that the encyclopedia would have to be free from all scientific jargon and would have to be written in such a way as to appeal to the intelligence of the average man." The memorandum shrewdly concluded that this approach "would also ensure a much wider sale than would otherwise be possible."[28]

In over 10,000 pages, Johnson attempted to address the various audiences indicated by these goals and presided over the creation of a document unique in American intellectual life: a monument to professional achievement, an accumulation of recent findings, and a guide for the citizen. While he had "a pretty free hand in the choice of writers," as he wrote one acquaintance, Johnson also had to define Seligman's rhetorical promises in his own terms. He did so, leaning toward readability and reformism. Declaring that "the mandarins are going to be conspicuously absent," Johnson nevertheless produced a finished work with a diverse and distinguished editorial cast. The somewhat deceptively named *International Encyclopedia of the Social Sciences*, in fact, has a much less cosmopolitan force of contributors than Johnson and Seligman had amassed a generation before. Technical articles abound in the *Encyclopaedia*, but social issues received far more attention than they would in 1968, when the successor volume appeared.[29]

Johnson, who did nearly all the final copyediting himself, maintained throughout the volumes a dedication to accessibility that appears to have irritated those who favored a more narrowly professional expository style. Seligman asserted in the preface that the editorial team sought to eliminate "all scientific jargon," but E. B. Wilson, the statistician who directed the SSRC, dissented from this stance. After examining the volume that included the introductory matter and ascertaining that Johnson wanted only "talky articles," Wilson feared that he "may have tried to be too intellectual and may have used too many formulas" in the article he was submitting at the time. He called "at least half" of the initial offering "pure piffle," and the conflict apparently was never resolved: no article in the *Encyclopaedia* bears Wilson's name. Other partisans of professional grandiloquence probably agreed.[30]

Johnson also despaired over his talented but overworked staff, pushing it nearly as hard as he pushed himself. Contributors, paid two cents a word with a five dollar minimum, flooded the editorial office with long, obtuse, and frequently unusable pieces, some requiring translation and most needing rewriting. The editorial pace of up to 45,000 words a week of final copy taxed Johnson's staff, yet he shaped them into a loyal and committed band. Despite telling the editor of the *International Encyclopedia of the Social Sciences* in 1964 that "in seven years our opinions never once conflicted and never the slightest cloud obscured our personal friendship," Johnson had in fact responded bitterly to Seligman's exhortation to speed up the copy: "We can get out the third volume in December, in spite of your rather fantastic calculations; but we are considerably less likely to do so if you call me up from Lake Placid and offend me so deeply that it takes three days to get back to a reasonably level head again. To assure you once for all, the third volume will be out before January 1, or you will have my resignation in your hands." He continued by attacking the publishers: "I am not a child [he was fifty-six years old at the time] nor a servant, to be sued through a guardian, and can take no cognizance of complaints addressed to others than myself." To his credit, Seligman smoothed relations with a graceful letter and expressed his "honest opinion . . . of profound gratitude" to Johnson.[31]

Seligman, meanwhile, had his own mission to complete. Edmund E. Day asked in 1932, "What are your present ideas with regard to the financing of the balance of the Encyclopaedia's requirements? It looks to me as if you still had a very large problem on your hands along this line." Philanthropic managers had watched the values of their portfolios drop substantially since they first approved of the project. The original cost estimate of $625,000, meanwhile, almost doubled over the next seven years to $1.25 million. With the volumes appearing one at a time, the foundations occupied the awkward

position of being unable to terminate the project without the most embar-
rassing consequences, so they stressed cost-effectiveness. In light of the
substantial outside underwriting, Seligman was pressured to adopt a more
conventional relationship to the project, and on February 20, 1931, he removed himself from the position of earning royalties on it. The foundation
administrators suspected inefficiency in Johnson's offices, so in 1932 the
Rockefeller Foundation hired a consulting economist to investigate the enterprise before the philanthropies watched any more money swirl into the
vortex of unaccountability.[32]

The editorial operation at once impressed and dismayed the consultant,
Mark Jones. Salaries, he advised, were overly generous, and the staff did
take pay cuts in the midst of the Depression as a condition of the continued
foundation support. The biggest cash leak, it turned out, remunerated the
editor-in-chief. "Dr. Seligman is understood to be a man of independent
means," Jones wrote, with an apartment at 145 Central Park West. A sum of
$58,170, over 5 percent of the total expenditures as of late 1932, had paid
his salary. The "disproportionate and unwarranted" investment of his client's money bothered Jones, who noted that Seligman "does not make his
office [at the *Encyclopaedia* office] and appears to do but a limited amount
of active work on the project." (In 1990 managing editor Max Lerner recalled this to be the case: Seligman "didn't do a god-damned thing. He never
showed up except on 'royal' visits.") Jones acted as a liaison between the
parties and suggested to Seligman that he accept the professional prestige of
his position in lieu of further salary. Arguing that "he was responsible in
large part for the success of the Encyclopedia," Seligman wanted to continue to draw the salary even though he had told Johnson two years earlier that
"you have virtually saved me all hard work."[33]

Jones saw through this sort of bluster: he praised Johnson for his dedication even as he chided the avaricious Seligman. "The greatest single contribution to the project has been made by Dr. Alvin Johnson," the report concluded. The production staff reacted to him by becoming a "loyal, devoted,
and competent group." (Lerner admiringly called him "the last man who
knew everything.") At the same time, Johnson had "carried on the work
mainly on the basis of personal ability and prodigious effort"; ironically,
given the encyclopedia's theoretical fondness for social planning, there appeared "to have been but minor resort to organization." Jones found this systemic inefficiency very difficult to eradicate; in his autobiography Johnson
recalled that the consultant could not even accuse the editorial team of wast-
ing paper clips. Eventually the foundations prevailed on the salary issue and

loaned the *Encyclopaedia* the necessary funds (instead of making further cash grants), and the appearance of volume 15 completed the set in 1935.[34]

....

The *Encyclopaedia of the Social Sciences* can best be understood in terms of its internal contrasts; many of the goals are if not mutually exclusive at least difficult to reconcile. A civic pragmatism moved the writers and editors to reach citizens with technically competent but carefully written entries on everyday topics. Professional aspiration, on the other hand, inspired methodological sophistication and technical specificity—in short, more jargon. Johnson worked demonically to make the writing both precise and readable, and the set achieved this goal spectacularly. As a distillation of the achievements of a generation, the *Encyclopaedia* assembled often impressive statements by the major minds of the era: Benedict, Boas, Malinowski, and Mead; Frankfurter and Pound; Keynes, Mitchell, Berle, and Means; Croce, Laski, Mannheim, Bloch, and Sombart; Sullivan, Thorndike, and Parsons; H. Richard Niebuhr, W. E. B. Du Bois, and Mumford; and Beard, Becker, and Curti. The *Encyclopaedia*'s utility as a reference tool forced libraries to rebind their sets repeatedly. (One study tried to quantify the use of different entries by rating the finger smudges on various pages.)[35]

The many positive reviews of the work frequently mentioned its implications for a democratic citizenry. The *New York Times* predicted that the set "will also be an invaluable source of information for the general reader," while *Political Quarterly* called it "a monument to reason in its application" in a time when the fascist countries had "banished . . . [all] rational thought." Instead, those countries were victims of "a monstrous collection of fantastic dogmas, grotesque myths and absurd legends." The publicity statement for the *Encyclopaedia* calmly claimed that "the Encyclopaedia provides full information on all questions relating to social progress." Lengthening shadows of what would soon be called totalitarianism thus demonstrated the desirability, nay, the necessity of more nearly consensual political reform proceeding from agreement on the facts.[36]

In contrast to the later *International Encyclopedia of the Social Sciences*, humanists exhibiting a frank antiscientism appeared in the *Encyclopaedia*. Neo-Comtean positivism was decried in the entry on the topic, which instead called for "philosophic reflection" as the guide to "reforming and reconstructive activity." Harold Laski, in "Bureaucracy," cataloged the many dangers of professional administrators. Elsewhere, art, not technology, was argued to be "reason and intelligence in operation." The philosopher Horace

Kallen invoked the spirit of William James when he claimed that "statisticians' correlations are sought as surrogates for the operative causes" in modern social science. For this "abstractionism" the "Jamesian attitude" provided a "salutary corrective."[37]

The editors themselves seemed at times to be critical of their primary constituencies. Only a multiplication of textbooks gave sociology "the appearance of content" in the interwar years, according to Lerner's unsigned overview of the social sciences. Political science was found to lack "either a clearly delimited set of problems or a definitely prescribed methodology"; practitioners remained unclear as to what constituted either science or politics. Aspirations toward mastering social process in a manner analogous to natural science, "variously known as 'social control' and 'social engineering,'" had "turned practically every discipline in the direction of the application of its precepts." The article warned against the deeper philosophic movement afoot. "In their attempt to escape the simplicity of unitary answers and yet avoid metaphysics," students of society embraced "a set of beliefs and techniques which were at once less demanding and more satisfying": pragmatism, quantification, and behaviorism. Such self-critical assertions indicate a fundamental ambivalence that appeared in only a few articles, those relating to social control, scientific management, and applied science.[38]

Within Johnson's larger attempt to define modern liberalism, the belief in an applied science of society appears frequently. The attempt to bring reason to bear on social problems grew by the 1930s to become a much more widespread call by self-proclaimed liberals for overarching plans, governmental efficiency, and expert direction. Johnson recruited leading advocates of applied scientific insight, including the longtime Taylor Society managing director Harlow Person, the *New Republic* editor George Soule, and Mary van Kleeck. Rexford Tugwell contributed after he had visited Russia in 1927; Rex the Red, as he came to be called by his many detractors, subsequently became the most vociferous proponent of centralized planning in the Roosevelt circle. From the ranks of government and personnel administrative reformers came W. F. Willoughby and Ordway Tead, respectively. Person and Soule together contributed a total of fifteen articles to the venture, including some on major topics. In addition, as a contrast to the perceived disarray of American responses to the Depression, "the Russian experiment has contributed enormously in making vivid the possibility of a deliberate change in any country in both the instruments and the aim of control," as one article put it. Notably, no entries were devoted to depressions, recessions, or panics.[39]

The argument that recurs in the scientific and managerial articles—not in most entries—became a staple of liberal thought: application of methods and of a spirit appropriated from the realm of natural science would enable modern democracy to achieve economic planning and social control. The Depression appears to have inspired an urgency that seems at odds with the calm rationality embodied in the mythical engineer implicit in scientism. Tead's definition of personnel administration, prescriptive and judgmental, provides but one example. It was, he wrote, "the planning, supervision, direction and coordination of those activities of an organization which contribute to realizing its defined purposes with a minimum of human effort and friction, with an animating spirit of cooperation and with proper regard for the genuine well being of all members of the organization." In a document celebrating an "objective" science of society, this entry swells with undefined value declarations and a resolute disregard of the possibility of conflict in the competition for limited goods. The convergence of scientific investigation, technocratic reform, and economic managerialism made for a compelling but tense reconciliation between social usefulness and scholarly objectivity.[40]

Morris R. Cohen's entry, "Method, Scientific," provides the clearest and most subtle discussion of the possibilities of science as a social cure. Avoiding the hyperbolic, he steered the discussion away from "the content of [science's] specific conclusions" into a consideration of method and spirit by which "findings are made and constantly corrected." This experimentalism overlaps Deweyan pragmatism, a common influence within the *Encyclopaedia*, but his sophistication made Cohen an exception; many other entries mimicked natural scientific discourse while maintaining an ill-defined faith in progress. "Science and technology," asserted the "Science" entry, "make it possible—if moral and practical development can keep pace—for men to realize the kingdom of God here on earth." One author defined social control as "active intelligent guidance of social processes" and found such an idea "thoroughly characteristic of the twentieth century." The engineer provided a model for the new social administrator, as Person, a longtime acquaintance of Johnson, explained: "Engineering as a technique has passed beyond the sole possession of the engineer and is now available to the economist, to other social scientists," and to business managers and owners.[41]

Such celebrations of social control—a nebulous concept rarely spelled out—only implied the limitations upon freedom necessary for such conceptions to be realized. The very advocacy in which the ostensibly objective reference work engaged further testified to the unclear boundaries within which the scientific reformers operated. The creators of the work addressed important concerns, however, as the undeniable complexity of a technological civilization demanded more from government than good intentions. Philanthropic managers and academic professionals thus searched for a method both effective and capable of instilling something akin to virtue. By turning to scientism they sought confidence in means while leaving overall direction an open question. At the same time, their prescriptions for social control were broad and moralistic, as the hard-core behaviorism of the *International Encyclopedia of the Social Sciences* demonstrates by way of contrast. The creators of the *Encyclopaedia* exhibited a hopefulness that no longer resonates; technology and science seldom appear unquestionably positive and no longer inspire political reform.[42]

The benefit of such infiltration of applied scientific method and spirit would ostensibly be in society, but professional social scientists saw how they would profit as well. "The social sciences are now reaching that degree of maturity which will permit them to be characterized as sciences, and in their professionalized aspects as applied sciences," proclaimed Luther Lee Bernard. The incipient prestige of these scientifically validated professionals depended on a market of citizens respectful of their abilities and cognizant of the expert's true station. Education thus would complete the transition to prestige and responsibility. Person found "especially necessary" the "training of an understanding, cooperative body of citizens." Bernard agreed. The problem, he wrote, was one of "procuring as wide a dissemination as possible of this [social science] research among the masses." The *Encyclopaedia* itself was of course one attempt to do so.[43]

How would specialized technical expertise harmonize with an educated democracy? In certain entries the *Encyclopaedia* tried to have it both ways. Initially, an entry devoted to the "Amateur" consoled the reader. The amateur is not the opposite of an expert, but rather the expert's "complement." Amateurs are freer to innovate because they are not bound by the traditions that define expertise. The expert, meanwhile, merely fills a need presumably defined by those amateur masses: "The use of the expert in government and social affairs has assumed prominence concurrently with the adoption of a new notion of the function of government itself." The passive verb leaves responsibility an open question: "This has come to be thought of . . . as the facilitation of the good life and the removal of friction by a technique of social engineering." Meanwhile the reader finds that expertise can redefine moral discourse: "Questions which political experts are called upon to decide, although ethical when regarded as ends, may be technical when regarded as means." The escape—from politics, from debate and compromise, from right and wrong—implied by these entries turned on an idealized version of science at odds with actual practice; no article provides any examples from real scientific experience beyond facile metaphors. But illusion had much more utility than mundane and ambiguous truth, so the illusion was pragmatically accepted.[44]

Within the *Encyclopaedia* itself, methods were frequently mistaken for goals, and the goal for many contributors was a rational society. Glacial conceptions of historical inevitability overshadowed the calls for social order: "The direction of modern society would seem to be toward a planned economy," read the entry on national economic planning. The economist Walton Hamilton claimed that "all that a people can do is shape as intelligently as they can a change which is inevitable." Such neo-Comtean notions, relying as they do on disembodied historical causation, seemingly contradict the control theme implicit throughout the *Encyclopaedia*. George Soule of the *New Republic*, however, put more responsibility on human agency when he summed up the opinions of some experts: few of them believed "that stabilization is possible without essential modification of the economic and political orders." Whose intelligence, whose modifications, and whose version of stability matter? These questions dangled, unanswered.[45]

Peter Rutkoff and William B. Scott have recently argued that a civic pragmatism heavily indebted to John Dewey shaped Johnson's editorial work, and their position is well fortified. Alexander Goldenweiser and Horace Kallen, two reigning New School pragmatists, contributed significant theoretical articles. Dewey himself, "something of a hero" on the editorial staff, wrote some of his most lucid declarations of the pragmatic spirit and method for this forum. But Thorstein Veblen must be considered the other intellectual patron saint of the venture. Johnson brought a strong appreciation of this eclectic economist to the *Encyclopaedia*. He assigned himself the entry on Veblen and later planned to write a book on the theorist before deferring to Lerner. In the Depression, the writings of Veblen, now safely dead, underwent a rebirth as many readers began to take his critiques of pecuniary behavior more seriously.[46]

In the *Encyclopaedia*, Deweyan experimental pragmatism and Veblenian social engineering ultimately converge on several key points even while each serves to highlight other distinctions between the two. Some articles in the

Encyclopaedia attempted to reconcile competing claims of value and fact, and of participation and expertise. The "Control, Social," entry concluded that "society must have prophets, poets and artists to give a vivid sense of new values and a host of economists, engineers and technicians who will translate those values into specific measures." Even so, the most innovative attempts at reconciliation came from the pragmatists Dewey and Kallen, and from Cohen, who refused the pragmatist label while sharing certain aspects of the outlook. Maintaining the spirit of scientific investigation and application, they sought a third path between mechanism and chaos with a series of long and provocative articles, especially Cohen in "Method, Scientific," Dewey in "Human Nature" and "Logic," and Kallen in "Pragmatism."[47]

Of the three, only Dewey expressed much confidence at the realistic possibility of a pragmatic zeitgeist. Cohen limited his claim to stating that the scientific method "does enable large numbers to walk with surer step." People can develop "policies of action and of moral judgment," responding to more than "immediate physical stimulus or organic response." The latter blast at behaviorism clearly indicates some faith in human possibility, but certainty, especially in light of recent theories by Einstein and Heisenberg with regard to nature's randomness, could not be promised. Kallen also affirmed the discomfort occasioned by a reliance on pragmatic investigation. While it merged "insight into scientific method with the mood of democratic social experience," pragmatism remained "too difficult a rule of conduct for many to live by." For men and women who invent social dogma because they cannot face "chance, change and process," pragmatism "demands too complete a disillusion." Kallen did hold open the possibility of more authentic human existence if the requisite self-confidence could be instilled, perhaps by the success of the applied scientific method.[48]

Dewey arranged his version of the argument on similar grounds, calling the split between fact and value "tragic." Humanity's "supposed ideals and aims have no intrinsic connection" to the means of their implementation, while "factual data are piled up with no sense of their bearing" on social policy. The method of scientific inquiry, and more necessarily its spirit, would lead individuals and societies to an appreciation of contingency that would allow true innovation and originality in the arrangement of social relations. From his fellow citizens, Dewey ambitiously hoped for "a willingness to substitute special concrete plans of modification for wholesale claims and denials; the growth of a scientific attitude which will weaken the force of ideas and battle cries coming from the past; willingness to see social experiments tried without interference by outside force; and the use of educational means that are regulated by intelligent foresight and planning instead of by routine and tradition."[49]

Such a scenario ignores the essential nature of politics as the necessary art (not science) of contestation and compromise. Where does the power originate to implement one plan and not another? Whose definition of intelligence carries the day? Foresight is admirable, but who decides the goal toward which it is directed? Who experiments on whom? If Dewey is so keen on moving scientific thinking into society, why does he dispense with "tradition" when scientific inquiry is fundamentally premised on an institutionalization of past mistakes, insights, and hunches? How can social thought become more scientific without inclusion of past results in the working lexicon of current investigation?

Perhaps more than most American intellectuals, Dewey presents his readers with a moving target, and to pin him down to this statement would misrepresent the issue. But for all his humane qualities, Dewey ultimately fell victim, in this setting, to the need for mechanistic reason that similarly marred the social thought of Thorstein Veblen. Science applied to society becomes the goal in and of itself. Dewey comes down on the side of social control—and in his world, a benign and liberating control it would no doubt be, as his intellectual biographer Robert Westbrook has argued—because the inherent logic of social engineering ignores the distinction between raw natural material and human culture. Social science, in this view, fails not because it homogenizes human diversity in order to fit human societies into mathematical or behaviorist constructs, but because it is not engineering on a par with the engineering of nature. Dewey bemoaned "remoteness of social method from guidance of social, legal and economic phenomena." He lamented "the failure to find a generally accepted method which will do in control of social forces what scientific method has accomplished in control of physical energies." He sought to make "both concepts and facts elements in and instruments of intelligently controlled action." The success of the radio designers, dam builders, and locomotive makers depended on authority, consistency, and imposition of will. How that success could be duplicated in human societies without an abandonment of democracy, pluralism, and freedom Dewey did not say.[50]

The *Encyclopaedia* as a whole frequently gets caught in a similar contradiction; Rutkoff and Scott write that Johnson and the *Encyclopaedia* were committed to "the twin authorities of scientific inquiry and liberal democracy." This particular scientific perspective reflects only the activist point of view in which inquiry is validated in results. That viewpoint, of course, is precisely the one Edmund Day shared with some of the social scientists the Rockefeller Foundation funded; he had said that "the ultimate test of scientific work in the social field must be pragmatic." Such pragmatic scientism, or scientific pragmatism, partially negates a commitment to democracy; the intellectual reliance on expertise implicit in social engineering make the mass gains in liberty and equality promised by the liberal code extremely problematic. Furthermore, Johnson and the editors failed to relate the ideals of liberal democracy to real-world political processes in any significant way, in part because the scientism served as a substitute for politics.[51]

In contrast to the many entries on managerial and social engineering topics, "Politics" runs only two pages. The term, according to the article, "comprehends scheming." It "invites action and speech less intellectually honest and high minded than are common in purely personal relationships." The author quoted George Bernard Shaw, calling the dramatist's definition "cynically put but profoundly true": the politician, in Shaw's words, "has now to learn how to fascinate, amuse, coax, humbug, frighten or otherwise strike the fancy of the electorate." (The *Oxford English Dictionary* also credits Shaw with the first deprecatory use of the word *scientism* in 1921.) A much more thorough entry, "Power, Political," ends with a warning. "The more astute dictator," wrote Hermann Heller, could draw upon technological innovations "to manipulate the manifold instruments of mass appeal and mass exploitation and thereby to achieve a monopolization of political power hitherto undreamed of." This fear of authoritarian abuses of social engineering contributed to the discrediting of the notion in America between 1940 and 1970. It is no accident that Heller, writing from Madrid, saw the clouds of technocratic abuse before most Americans did.[52]

Philanthropies with a heavily vested interest in a particular variation of political economy initiated and supported reform programs built around presumably objective social scientific technicians and managers. (Nowhere, it should be added, could I find evidence of meddling by Day, Keppel, or Ruml. Most of the staff leaned to the political left, with several communists among the editors.[53]) At the same time, social scientists in this period encouraged education in and a reliance on the benefits of "objective" expertise, a program that would benefit them professionally. In such a scenario, just how open to the experimental evidence could the technicians be in the face of their own rational self-interest? The social matrix within which science, and scientists, function was taken as a given by many associated with the *Encyclopaedia*. But society is not a fait accompli, resulting instead from relationships of power and, more frequently in this period, knowledge. The fusion of the two—a union catalyzed by the spirit of scientism—occurred within philanthropies devoted to improving society under capitalism and within professions pursuing both intellectual and self-interested goals.

RECENT SOCIAL TRENDS

Soon after his election, Herbert Hoover initiated the most original attempt to link social scientists, philanthropic administrators, and political leaders in an apolitical reform project. He convinced the Rockefeller Foundation to support a huge investigation into American social conditions, a stock-taking from which rational progress could originate. Mitchell, Merriam, Ogburn, and Odum helped carry out the resulting project, which failed to introduce scientism as a reigning tenet of political administration. Instead, the Depression reinforced the primacy of political negotiation, moralistic preconceptions, and intangible emotions for governmental success even as the rational reformers sought to prove otherwise.

Because Hoover worked to motivate peacetime America with the same sort of noncoercive (in his mind) patriotism that had energized the war mobilization, his Department of Commerce nurtured semiofficial relationships linking trade associations, professional groups, and other private-sector entities with government bureaus. By differentiating between private power —of cartels, vertically integrated firms, or foundations—and the constitutionally circumscribed power of government, Hoover developed a unique political economy. In theory, perfect free-market relations depend upon perfect information, and Hoover had labored throughout the 1920s to remedy the shortage of dependable social facts from which rational decisions could originate. *Recent Social Trends* continued, and built upon, the precedent set by White House conferences, data-gathering by the Departments of Agriculture and Commerce, and *Recent Economic Changes*, the report of the 1921 Conference on Unemployment.[54]

Hoover's aide in charge of *Recent Social Trends*, French Strother, asked Odum to float the idea of such a study at the Hanover conference of 1929. Once involved, Ogburn immediately approved of the plan, and he, Odum, and Strother convinced Mitchell and Merriam to enlist. After the group— without Hoover—incorporated itself as the President's Research Committee on Social Trends, it applied to the Rockefeller Foundation for support, and a grant of $560,000 was approved in December 1929. Mitchell and Merriam became chairman and vice-chairman, respectively. Ogburn served as director of research, with Odum as assistant director. Shelby Harrison took on the dual role of secretary-treasurer. Hoover was represented by Edward Eyre Hunt, the executive secretary of the group. It is worth noting that in Hoover's original plan as presented to the Rockefeller Foundation leadership, the experts' findings would be supplemented by "large meetings" at which he hoped to "stir up interests and secure action throughout the country." Such meetings never took place; instead, conflicts of vision between the social scientists and Hoover became more and more pronounced.[55]

Eventual sparring over purposes, methods, and procedures resulted from the different conceptions of research and reform among the various constituencies. Harrison represented a social work background from which many academic social scientists diverged in the 1920s. Hoover and Hunt, meanwhile, retained notions from engineering, in the former case, and Taylorism, in the latter. Expert, impartial investigation could be reconciled with social change by distancing the investigators from political partisanship. Here the Rockefeller support made the venture more like the SSRC projects than a congressional inquiry. The resulting endeavor could thus more plausibly be called objective; in their self-conception the researchers were scientists, not political appointees, as the group's publications insisted repeatedly.[56]

A commitment to scientism by the investigators led to repeated squabbles with Hoover's aides, many of which the social scientists won. Strother, for example, wanted someone with a noteworthy appreciation of written English to take part in the editorial process, but even historians like Allen Nevins and Carl Becker—much less such unprofessional presences as Ray Stannard Baker, Stuart Chase, and Mark Sullivan—were refused. Hoover, meanwhile, repeatedly asked for summaries of current findings to help him understand and control the effects of the Depression. He too was rebuffed as the social science–philanthropy network instead stressed professional behavior as it had come to be understood by the SSRC, the NBER, and university research centers. After addressing the committee on uncomfortably "political" issues—the link between immigration and organized crime, for instance—Hoover inspired increasing wariness within the project leadership that became fractious in the election year 1932. Seeking to protect the writers from being asked to change their articles in midstream, Merriam objected to Ogburn's practice of submitting "piecemeal" and unfinished reports to the president. The possibility that committee materials would be part of

Fig. 7-2. President's Research Committee on Social Trends (Herbert Hoover Presidential Library)

Hoover's reelection campaign was rendered moot, however, as the editorial work continued too far into the campaign season to allow prepublication publicity to begin before autumn.

Despite their fealty to scientific professionalism, the committee members differed as to the concrete manifestations of their outlooks. How readable, for example, should the final document be? Was it to endorse academic terminology or citizen accessibility? The breakdown fell along predictable lines, given what had transpired within social science during the 1920s. Ogburn led the movement toward facts and facts alone. "The alternative of fact," read Ogburn's "Note on Method," "is opinion," which, in turn, was "statements, the support of which is not demonstrated by fact." Notwithstanding tautological reasoning and convoluted prose, both of which troubled some of his colleagues, Ogburn pressed on. "The staff were not expected," he lectured, "to give their opinions on what to do . . . but rather to present *knowledge* which might be used for better action and opinion." Ogburn denied that he wanted to address the findings to an elite audience, despite his continued insistence on technical vocabulary. He feared making false promises and so was concerned when "a substantial proportion of the chapters that have been turned in have been criticized . . . as having too optimistic a tone." Even the word *trends* proved problematic for Ogburn, who worried about the appearance of prediction in a scientific document: he ordered the staff to "keep projection of trends as free from wishful thinking as possible."[57]

In the opposing dugout stood Merriam and Mitchell, who, writing more fluidly, reflected a more intuitive grasp of reality than could be achieved within Ogburn's wooden categories. They worried about their readership among an unprofessional public both concerned with the Depression and hopeful that social science might bring relief. Mitchell's introduction to *Recent Social Trends*, meanwhile, found him writing in ways almost calculated to pique his overly literal colleague. Instead of facts, Mitchell invoked familiar metaphors of awkwardness and control. "It is almost as if the various functions of the body or the parts of an automobile were operating at unsynchronized speeds," he wrote. But Mitchell, as he had for years, did not ask social scientists to be social engineers, for the data simply were not ready. He claimed that though "economic planning is called for," the phrase represented "a social need rather than a social capacity." All that could be achieved was "to lay plans for making plans." Once at the NRPB under Franklin Roosevelt, Merriam and Mitchell would title their first major document "A Plan for Planning," maintaining the continuity of their careers. They pursued rational social control, but apparently only as a long-term prospect.[58]

The possibility that some formal policy entity might result from their experience inspired stealthy optimism among the social scientists who worked on Hoover's investigation. Self-importance lurked from the outset, as the academics tried to impress observers with the implications of their still-forming professional identity. Merriam incorporated large roles for social scientists in the closing chapter of *Recent Social Trends* on government and, like Hoover, looked to "quasi-government corporations" similar to trade associations that would include academics in policymaking. Odum hoped that the SSRC might absorb some of the functions of the committee as ongoing projects. Once it became clear that Hoover's doom was sealed, even his last supporter on the committee—Ogburn—made overtures to Roosevelt's associate Raymond Moley, asserting that the investigation would have considerable value to the incoming administration. Being dismissed by the new president on political grounds appears to have been a real fear among the investigators. The stance of objectivity, though, served them well, for several occupied significant positions in the New Deal once Hoover could no longer claim their disinterested allegiance.[59]

When the project neared completion in late 1932, a question arose: What catchy phrase could advertise the 1,500-page tome? In March, Odum had suggested *Social Trends in the Remaking of America*. "There is enough vividness in its title and enough dynamics in its content to appeal to a large body of American people who are ready to kick off on a new era," he explained to Ogburn. At a meeting of the committee in December, a discussion devoted to publicity revealed yet again incongruities between restrained objectivity and the excitement of relevance, between the aim for precision and the glamor of commerce. Hollywood, science fiction, and current technology all contributed motifs to the discussion. When catchphrases were solicited, Mitchell suggested *Social Diagnosis*, and Ogburn followed with *Bird's-eye View of the Panorama of Civilization* and *Legacy for Future Policy Makers*. Harry Venneman, an editor, thought *Basis for Social Planning* would appeal to readers accustomed to blueprints for systematic change. Strother tossed out *Scientific News Reel of Present-Day American Civilization* and *Changing Status of a Nation*. Merriam put forth *Filming American Civilization*, then Hunt concluded the discussion with *Yesterday/Today and Tomorrow of American Civilization* and *Snapshot of These Changing Times*. Even though *Recent Social Trends* contained studies of leisure, religion, the arts, and other topics, the engineered dimensions of the current cultural landscape clearly ranked highest.[60]

Response to the volume exemplified the difficulties inherent in distinguishing fact from opinion and investigation from action. The vast, dense, and often obscure contents of the finished work clearly swamped some reviewers. Others applauded. John Dewey, for example, claimed that "the facts are presented . . . to make *problems* stand out, and that, in my judgment, is the proper function of statements of facts." The economist Adolf A. Berle, however, took a more activist line, challenging the investigators' decision "to state facts, rather than to interpret them or plead a cause." Perhaps, he wondered, "the desire for objectivity has been carried entirely too far." Charles Beard, an intent observer of scientism's progress, speculated that the volume would mark the end of the simplistic empiricism that held that "when the 'data' have been assembled important conclusions will flow from observing them—conclusions akin in inevitability to those of physics or mathematics."[61]

Meanwhile, in a heated exchange in the *Journal of Political Economy*, Ogburn defended *Recent Social Trends* against three major criticisms. The sociologist Pitirim Sorokin challenged not the investigators but "the inadequate methodology and 'philosophy' on which the investigation rests." His critique centered on inappropriate quantification, fragmentation in place of organic interrelation, and "fact-transcribing" where fact finding had been promised. Ogburn's reply declared, in turn, that "the report does not go in for an intellectual display of ideas, values, analysis, and concepts as such"; that "what Professor Sorokin seems to want is a social philosophy"; and that quite a lot of data were in fact new. He then concluded cryptically that "the main and only real methodological difficulty of the study was not mentioned by the reviewer." Such resolute talking past each other—"Yes it is," "No it's not"—further illustrates how something apparently as straightforward as "reason" could be so contested and political.[62]

. . . .

Many similarities of personnel, funding, and objectives connect these three ventures. Merriam's SSRC provided logistical coordination for each undertaking. The *Encyclopaedia of the Social Sciences'* Alvin Johnson joined *Recent Social Trends* planning sessions, as did such dedicated social engineers as Hornell Hart, Paul U. Kellogg, and Stuart Rice. The man Merriam sarcastically called "papa Seligman" deigned to appear at Hanover; Johnson came as well. Edmund Day represented the Rockefeller presence in each project, solidifying his position as a key liaison between the social scientific and philanthropic reform communities in this period. Despite such close coordination of academia, foundations, and government, the fundamental limitations of the natural science model rendered each project unable to fulfill the promise of scientism. As Charles Beard asserted in his review of *Recent Social Trends*, politics meant "the perdurance of outside activities, forces, and powers."[63]

Each project required Rockefeller funding and, probably more important, Rockefeller legitimacy for its influence to be felt, regardless of the degree to which Day controlled or merely reflected the ideology of the venture. During these same years, other reformers devised similar projects. Unlike these three, they could not capture much popular enthusiasm, financial support, or institutional warrant. Why they failed, despite sharing important assumptions with the social scientists of this and the previous chapter, demonstrates the primacy of philanthropic managers for rational reform.

8

ROADS NOT TAKEN

In contrast to the generously funded social science ventures discussed previously, many attempts to engineer society amounted to little beyond hopeful grant applications. These lesser-known efforts do reveal, however, the degree to which rational reform connoted credibility and progress. A sociologist's ambitious educational project, a radical foundation administrator's international program, and some engineering reformers' attempt to return to the center of the movement they had helped create all shared a common fate. The major foundations declined their proposals, but these reformers, in their enthusiasm for various renderings of social engineering, testify to its appeal in pre–New Deal America.

A CENTURY OF PROGRESS, 1933

The most extensive plan to incorporate social science into a world's fair occurred in Chicago, home of both the monumental Columbian Exposition of 1893 and the University of Chicago's brand-new social science research building, a gift from the Rockefeller philanthropies. The forces embodied in the two artifacts failed to mesh comfortably, and the reasons why tell the observer much about the competing commitments entailed in the admiration of both quasi-educational pageantry and applied science. Under the direction of Howard Odum, the drive to install social science exhibits in the 1933 Century of Progress exhibition stalled out; someone regarded as a real scientist—Fay-Cooper Cole, an anthropologist from the University of Chicago—eventually accomplished the task. Odum's involvement illustrates the distinctions between natural and social science and the problems of bringing his vision to the public.

Organizers in Chicago had conceived of the entire fair as a testimony to the power of science. The "Century of Progress" that elapsed between Chicago's founding as a village and 1933 "represents also the great period of development of the physical sciences and their application to the services of man," read one of the earliest documents of the fair committee. They felt that the exposition should emphasize "the service of science to society and the benefit to humanity brought about by this scientific and industrial development." To be authoritative in the performance of this mission, the fair enlisted the services of the NRC, an umbrella group of natural scientists and engineers that dated back to the Civil War. During World War I, Woodrow Wilson reorganized it, and by 1928 the NRC had become the preeminent scientific body in the United States. On October 9 of that year the NRC officially associated itself with the fair and authorized the appointment of an advisory committee. Frank Jewett, the president of Bell Laboratories, headed the resulting body, which also included the scientific popularizer Michael Pupin and Max Mason, a former president of the University of Chicago who headed the Rockefeller Foundation after 1929.[1]

The union of the prestigious NRC with a world's fair resulted from a year of lobbying by Rufus Dawes, the Century of Progress president. He met with Pupin early in 1928 and convinced his committee colleague William A. Pusey, the retired head of the American Medical Association, of the importance of the NRC for the fair. Pusey then helped Dawes pitch the idea to other NRC members, who were in part convinced by Dawes's explicit linkage of industrial commerce and scientific research. The NRC, Dawes argued, stood for "the very thought we want to put into the fair, to wit: the advantages of a close alliance between men of science and men of capital." The NRC apparently approved of such an alliance, for it lent extensive assistance: Jewett's Science Advisory Committee eventually consisted of thirty-four subcommittees made up of over four hundred NRC members. These bodies gave Dawes and his staff detailed recommendations on exhibits in addition to constructing an overall ideology for the fair, which Pupin called "scientific idealism." It quickly caused problems for social scientists who wanted to be included.[2]

The NRC advisors could not define *science* in any standard way, as the transcripts of the Science Advisory Committee meetings show. Some of the group's romanticization and sentimental personification remain embarrassingly silly. Pupin was usually a prime offender. For instance, he "very much opposed" the "many [scientific men who] are very fond of saying that we, in our fight with nature, force her to reveal her secrets." Instead, Pupin favored the "idea that we should present allegorically science as a loving mother to whom man is listening." Maurice Holland, the NRC executive secretary, agreed. "Tom, Dick and Harry and his brother and sister will be able to actually meet Science [at the Century of Progress]." Such familiarity did not imply that "she [Science] will be dressed gaudily, cheaply and will become a sensational hussy merely to make a hit." On the contrary: "Nothing of the sort. Science will be shown as she really is—a friend to everybody, as simple in her ways as an old shoe." Such bizarre invocations of the whore/goddess dichotomy in the effort to democratize science came in response to a wave of realization that the complications of scientific innovation could outpace humanity's capacity to control it. Accordingly, when Odum proposed that industrial applications of science "in turn develop social 'problems,' which in turn give rise to the need for social engineering" from social scientists, he directly challenged one of the fundamental precepts of the enterprise.[3]

The iconography of the fair tightly coupled science and capitalism, both portrayed as products of rigid causal laws. As the official guidebook explained, "You trace the economic aspects of industry, of agriculture, and see . . . the reasons for the prices of things, the cost of making, and the profit." Prospective exhibitors heard in concrete terms how science would help to justify existing political economy: "Dramatic attention-compelling exhibits at A Century of Progress, influencing the public logically and favorably, will add enormously to the integrity of our *industrial and financial structures* and . . . not sacrifice the unprecedented advantages drawn from these institutions." Indeed, the political aspect of the fair would seem to be tailored to fit the social engineering outlook, but the agents of influence in the fair's metaphysics were not sociologists or economists. Because "so few of us realize that in virtually everything we do we enjoy a gift of Science," the public needed education in the real engines of progress: "They are the forces of science, linked with the forces of industry." It followed, in the words of the official motto of the fair, that "Science Finds—Industry Applies—Man Conforms." In every aspect the Century of Progress celebrated technological determinism.[4]

The scientists of the Science Advisory Committee most willingly reinforced the connections of natural science and engineering to society, but to use the name of science for academic investigations of those connections was blasphemy. At a meeting of the advisory committee, Edward Huntington told his colleagues that many of his mathematics subcommittee members would never have volunteered without understanding "that the main purpose of the fair was to dramatize science in the sense in which the word 'science' is commonly used by the man in the street, namely sciences included in the National Research Council." He thought that the public "will be very much confused if they find included under the name 'science' a great many things that they didn't suppose were included under the name 'science.'" Foremost among these, Huntington said, was "social engineering." Jewett quieted the objections by downplaying the issue of social science representation. Cole, the eventual director of social science exhibits, affirmed that "I don't think we are considering putting [Odum's exhibits] in the main hall of science."[5]

That building, initially called the Temple of Science, constituted the architectural focal point of the fair (see fig. 8-1). In the end the social sciences did not get space there; instead, a section at the northern end of the Electrical Building was set aside to give "some hint, at least, of the effect of this [scientific] progress on man and society," in the words of the fair's final report. In every regard, the organizers marginalized social science at the Chicago fair. They reconstructed a Mayan temple, and a "small, but interesting series, of Social Science Exhibits was gathered." Ultimately, the Century of Progress had nothing of the scale, drama, or impact that Odum craved. He had come to the fair office on leave from the University of North Carolina with his customary burst of energy. As the president of the SSRC wrote one associate in 1930, "Odum is just chock full of the Fair. He talks of nothing else." But his inability to enlist foundations, to overcome scientists' suspicion, and to attract exhibitors contributed to the failure of his vision.[6]

Odum devised an elaborate organizational scheme for the social science exhibits that involved many professional leaders and important members of the SSRC: Charles Merriam, Robert Crane, Arthur Schlesinger, Robert Lynd, Mary van Kleeck, and Shelby Harrison. He sought to display some findings of the Recent Social Trends project and looked to civic and professional associations to contribute materials on home ownership, child development, and adult education. His sense of mission directly resulted from the opportunity the fair presented: social science could stand with the natural sciences in the public imagination, sharing in the prestige of the scientific method and its contribution to progress. A memo of late 1931 both outlined the challenge and suggested the degree of resistance Odum encountered. "The Exposition," he wrote, "presents both an opportunity and obligation for the Social Sciences to demonstrate for the first time the progress of Social Science and Social Research and their application to human interests and human welfare." He wondered, however, "whether they have sufficient imag-

Fig. 8-1. Hall of Science, A Century of Progress (A Century of Progress Records, University Library, University of Illinois, Chicago)

ination and administrative ability to tell their story to the world." The realities of fair administration quickly became burdensome as Odum discovered his embattled position and the lack of any allies of consequence.[7]

In the idealized form in which Odum imported it into social science, science was objective—abstracted from both history and current passions —and efficacious. As it was understood by the fair backers and the NRC advisors, meanwhile, science was alternately comfortably familiar and formidably powerful. In addition, the intertwining of science, engineering, and industry gave science a political cast that was undeniably capitalist. As the historian Robert Rydell notes, "The science exhibits were intended to exemplify 'the idea of scientific and industrial unity' and to inject '*system* and *order*' into the exposition and, by extension, into American culture as a whole." By being so closely identified with corporate industry and the status quo, science became more distant from even the mild degree of social change advocated by the SSRC and similar groups.[8]

Odum tried in vain to duplicate the natural scientists' linkage of objectivity and commercial attractiveness. He declared, perhaps in a moment of wish fulfillment that would be neither his first nor last, that the social sciences "have attained scientific proportions in their methods, results, and in the rating which they hold in the worlds of education and practical affairs." How would such weighty enterprises be portrayed to the public? "Showmanship" became a primary consideration in the organization of the social science exhibit insofar as "the public will wish to have variety, movement, expansiveness, and vividness, together with facilities for coming together and for a certain amount of leisure time, contemplation, and entertainment." Odum attempted this Disneyesque task by asking for an SSRC advisory council and soliciting funds for a large and entertaining exhibit. The latter task proved to be impossible, especially in the midst of economic recession.[9]

Unfortunately for Odum his advisors brought to their task little of the zeal that characterized the NRC board in the natural sciences. Nevertheless, he pleaded for social science involvement because the fair mattered both to the public and to the practitioners. "If, from the practical standpoint," he argued, "human engineering is ever going to become as important as technological engineering, the findings and methods of the social sciences must be made known to the general public." Public displays like the fair "furnish an ideal opportunity for putting graphically before the people what we have been grinding out in our social science laboratories." Odum also tried to persuade social investigators that the exhibit would prepare fairgoers for the prospect of more research. Because "human problems can only be investigated with a maximum of cooperation and understanding," the public needed to know what social scientists were doing. "If we wish to get the access we need to personal documents, business documents, governmental records, et cetera," the "subject" of social research needed to understand the field. Therefore, Odum concluded, "the people, whom it is our function as social scientists to study, and as social engineers to help, should have as clear an understanding as possible of our aims and our results." Neither reason, however, compelled social scientists to action.[10]

In 1932, after Odum had stepped down, the SSRC declined Cole's request that they sell $100,000 in world's fair bonds and instead resolved, lukewarmly, that "any World's Fair that fails to include a substantial exhibit of the development of the social sciences will be seriously defective." The diffidence of the SSRC response to Odum contrasts sharply with the vigor of the NRC involvement and suggests that even social scientists themselves doubted their public importance or claims to scientific validity. The luckless

Odum did get some help from his Recent Social Trends colleague William Ogburn, but few other allies materialized.[11]

Professional concerns aside, money remained Odum's major headache. Even as he was just beginning his work, fair administrators worried about the financial viability of social science exhibits, but how much they told Odum is unclear. "Unless I mistake trends, it is going to be hard sledding to get cash to put on these [social science] exhibits," wrote the Century of Progress staff member Rudolph Clemen, who hoped that industry would contribute. "For example," Clemen continued, "any one with any knowledge in the field of economics, would feel that we ought to have exhibits in the field of money and banking, insurance and advertising." Such obviously commercial prospects juxtaposed sharply with social scientists' self-perception; Clemen and Odum appear to have clashed repeatedly. E. B. Wilson suggested getting foundation help from Rockefeller or Carnegie for the exhibits, but no aid materialized. Odum had to try to put on a fair without benefactors.[12]

The struggle quickly enervated Odum, and about a year after assuming his post, he quit in December 1931. His letter of resignation implied that his superiors were not taking the social science exhibits as seriously as Odum desired: "I beg again," he wrote the director of exhibits, "to urge the importance of making the Social Science exhibit a major unit of the Exposition." The plaintive tone belies a sense of desperation; Odum stepped down only eighteen months before the fair opened. A week later he wrote an unaddressed memo, perhaps to himself, outlining the "obstacles" that stood between him and an acceptable social science display. He had few supplemental funds; the space allocated to the social sciences was both small and undesirable; and those groups who could be included refused to purchase space in the social science display—organizations with funds wanted to exhibit in the more glamorous buildings. Social science's distinct lack of public appeal defeated Odum; in isolation its boosters glorified their field much more than the fair organizers, natural scientists, or business groups thought fit.[13]

The specter of natural science hegemony persisted, for even after Odum left Chicago the SSRC still worried about the second-class status of their "sciences." E. B. Wilson wrote Robert Crane of the SSRC in early 1932 that "the present problems concern the interweaving of these [read "our"] plans with that part of the natural science exhibits which is most distinctly social, namely the part on psychology and anthropology." Such definitional disagreement was inevitable, for the NRC accepted only those two fields as sciences. After Cole took over for Odum, he relegated sociology, economics, and political science—the core SSRC disciplines—to their ignominy. In the fair guidebook, social science is presented nearly as a branch of ethics: "Perhaps you will find an answer to the perplexities of the present that cause our sometimes querulous questioning of the worthwhileness of things." The Mayan temple (fig. 8-2), meanwhile, enshrined "scientific" anthropology as much as any Yucatan deity. Native Americans displayed as though in a zoo reinforced the progress theme through their "primitive" contrast to the laboratories and machinery at the heart of the fair.[14]

After Odum left Chicago, he threw himself into other projects, including a regional survey of the South under Rockefeller and SSRC auspices. He did write his administrator at the fair, telling her that "that sojourn was both a delightful and disappointing experience. I still have many waking moments wishing we might have attempted the picture in its entirety." If, as Daniel Singal suggests, Odum exhibited "basic ineptitude in the art of grantsmanship," the Century of Progress could well have been a painful education. Rufus Dawes, the president of the fair corporation, was an experienced businessman and skilled organizer, as he showed in his courtship of the NRC. The fair manager, Lenox Lohr, appears to have been similarly adept, for he managed to get the exposition to turn a profit in the midst of the Depression. The NRC science advisors came from the highest levels of industry and commerce—Bell Laboratories, the Rockefeller Foundation, the Rockefeller Institute for Medical Research, Columbia University, and a leading New York construction firm. In this setting, the bluster, enthusiasm, and self-perceived connectedness that had helped build an empire in Chapel Hill were probably swamped soon after he accepted what appears to have been a dead-end job even in its conception.[15]

Various aspects of the Century of Progress illustrate the gulf that remained between natural and social science, despite the frequent claims among social scientists of their newfound authority. The funding agencies that supported such projects as the Recent Social Trends investigation shied away from the commercialism and superficiality of a world's fair as a venue for public education, so Odum had to fight on behalf of a lightly regarded constituency for both intellectual and financial existence. That he was unable to bring sociology, social work, and political science to the fair indicates the unwillingness of many Americans to underwrite social engineering in meaningful ways. The movement began on the margins and never really left them, despite the participants' aspirations and self-importance.

Fig. 8-2. Mayan Temple, A Century of Progress (A Century of Progress Records, University Library, University of Illinois, Chicago)

MARY VAN KLEECK

Behind its impressive name, the International Industrial Relations Institute was essentially a two-woman transoceanic operation. Mary van Kleeck (1883–1972), the American half of the pair, provided the rational reform movement with an ongoing, distinctive, and challenging voice. After working in the mainstream of Taylorism for well over a decade, she began in the IRI to turn her unique merger of engineering-based managerial technique with social work's concern for the downtrodden in a new and radical direction. In the decade after 1925, van Kleeck came openly to advocate Soviet-style socialism and was eventually rewarded with a subpoena to appear before the McCarthy subcommittee in 1953.

After receiving a B.A. from Smith in 1904, van Kleeck began her work as an industrial investigator and soon learned the ways of the New York social work community. She studied with such important figures as Edward T. Devine and Franklin H. Giddings while also becoming involved with the

Fig. 8-3. Mary van Kleeck (Courtesy of Rockefeller Archive Center, Pocantico Hills, N.Y.)

New York Women's Trade Union League. The newly formed Russell Sage Foundation began to fund her research in 1908, and she was promoted to director of industrial studies in 1910. Except for her war work of 1918–19, van Kleeck stayed with the Sage Foundation until 1948. She sat on the 1921 Unemployment Conference with Ida Tarbell, Wesley Mitchell, and Owen Young of General Electric, among others, and on the 1922–23 Committee on Unemployment and Business Cycles. A close associate of Morris Cooke, van Kleeck was one of the few women to be admitted to the Taylor Society, a group with which she maintained contact. She initially supported Herbert Hoover's associational model of economic coordination in the early 1920s before becoming disillusioned. After van Kleeck's withdrawal from these circles the IRI took on added importance, especially as the Depression further discredited Hoover.[16]

Like other Taylorites, especially Harlow Person (who sat with Cooke on the IRI board), van Kleeck logically concluded that the business firm was not the only social structure in need of rationalized procedures and relations. She wrote Cooke that "the lack of balance between human needs and the resources which actually exist for meeting them . . . suggests the necessity for applying over a wider scale the methods of research and planning which the scientific management movement has introduced." Van Kleeck extended the scope of Taylor's theory and called for national and international planning as the solution for the problems her social work investigations revealed; the "single objective" of such planning was to use resources "in the service of all the people of the world." In case Cooke doubted, she added, "this is not Utopian, but a practical, realistic application of the methods of science." Her mixture of social work and engineering had at its center the enhanced well-being of the worker, but systemic change, rather than local and temporary palliatives, was her goal. Modified managerial innovation would result in steady employment, higher wages, and lower-priced mass-produced consumer goods, rather than in higher profits for business. This notion obviously represented a refinement and integration of ideas first offered by such social technicians as Simon Patten, Thorstein Veblen, and E. A. Filene.[17]

The IRI was founded in the early 1920s by personnel managers, many of whom were women, who had seen in the wartime expansion of the work force a need to rethink labor-management relations. After an initial meeting in 1922 in Normandy the reformers reassembled every three years to consider issues related to worker welfare, industrial management, and larger social changes. Just before the 1925 conference in Holland, the group changed its name to the none-too-pithy International Association for the Study and Improvement of Human Relations in Industry. More than fifty delegates, still mostly women in social welfare positions, represented over twenty nations. After the success of the 1925 meeting, shorter summer sessions were scheduled the next two years before the triennial meeting in 1928 at Cambridge, England. By this time, despite the growing rosters at the meetings and expanded publication of the proceedings, van Kleeck and her Dutch colleague Mary Fledderus had become the essence of the IRI, helping to keep it alive by securing financial support from the Sage Foundation.[18]

After the onset of the Depression on both sides of the Atlantic, the 1931 conference in Amsterdam attracted wider attention, for planning had quickly become a topic of popular discussion and not merely the concern of specialized and dissident technicians. The IRI organizers seized on this inquisitiveness and sought to establish a world social economic center. Van Kleeck's description of the center's planning possibilities drew upon a most powerful—and familiar—language of applied scientific thinking. "When the engineer studies how to build a bridge," she told the Amsterdam congress, "he takes over scientific discoveries and applies them to building a bridge which will carry the traffic. Those who are responsible for business and industry today . . . must also learn from economics and scientific management the methods of attaining a desired end." In search of appreciable political power, van Kleeck turned to applied science for her illustrations.[19]

Van Kleeck had tried to interest Edmund E. Day and Raymond Fosdick in a center that would address two sides of "social economic planning": population and economic resources. As the Amsterdam conference program asked, "Can the methods of science be utilized to achieve balance between resources, production and consumption? Can science be substituted for casualism in the development of economic policy?" Probably because of the group's politics, the Rockefeller Foundation declined to fund such a center even though the proposal used all the right phrases.[20]

The IRI pointedly opposed the hierarchical management that Taylor implemented and Day favored. According to van Kleeck's closing statement to the 1931 meeting, "the immediate aim of political action needs to be redefined in terms of its social effects[,] and the technical procedures for achieving the end desired must be worked out by the technicians." But those technicians, no matter what their expertise, had to serve the citizenry. The "Congress of average citizens" at Amsterdam was established "to consider what may appear to be the exclusive work of experts and professionals" in order to educate the masses, and to enlighten technicians who "tend to work in separate compartments." In 1931, van Kleeck blurred the lines usually drawn between skilled technicians and unskilled laborers; by 1934 she would go so far as to argue that true technicians were workers and not allied with the middle-class industrial managers. "Workers' control is substituted for possessors' control," she argued in an updated version of the Veblenian scenario, "because workers' control is consistent with technological development and application of scientific knowledge." Even though Taylor had distrusted the profit motive, no system of managerialism had ever allied labor so closely with scientific managers in opposition to capital.[21]

Except one, in van Kleeck's view: the Soviet Union. After the officials behind the Five Year plans were invited to Amsterdam to discuss government control, van Kleeck appears to have stiffened in her conviction that such centralized application of scientific expertise portended a more humane future. She traveled to the Soviet Union in 1932 and was excited by the prospects there; "power [of a despotic sort] gives place to the power of knowledge," she wrote. After returning, van Kleeck resolutely defended any and all Soviet actions but apparently did not take the step of officially joining the Communist party. As one interviewer wrote after speaking with van Kleeck, "I started to criticize their ruthless methods . . . but she matched each one with a deed yet more ruthless here in America—massacres in Colorado, . . . in North Carolina, lynchings, the denial of all our liberties in times of stress." This spirit of determination, her critical examination of American social reality, and the pursuit of an apparently logical solution to pressing problems defined van Kleeck's career.[22]

Because the Sage administrator saw clearly the degree to which Taylorism oppressed workers by "de-skilling" them, her version of social engineering might well be titled socialized engineering, for she explicitly rejected the appropriation of scientific social knowledge by a cadre of experts. In fact, nearly alone among American rational reformers, she saw socialism as a logical program of centralized planning and administration. "Knowledge, in which all persons in an enterprise should have a share, is the master" in truly scientific management, she stated. On the other hand, "we think it is time to put the wisest in all countries to work upon this problem" of depressed economies. Such statements sound contradictory until one recalls van Kleeck's commitment to empirical education. Creating an attitude to support comprehensive planning fell to education, but in her social theory, "education is education through participation in the process itself." For this reason she criticized social workers, encouraging "a closer association with the workers' groups than with boards of directors and governmental officials." While her own position brought her into frequent contact with precisely those sorts of individuals, her radicalism balanced power in the boardroom with class consciousness.[23]

Moving beyond Taylor's promise of a "complete *mental* revolution," van Kleeck saw open class conflict as fully compatible with rational planning. "The delusion that human rights are attainable without struggle . . . is responsible for weak programs and ineffective activities," she told a group of social workers. Even so, reconciling Marx with managerialism sometimes could sound confusing. While "the working hypothesis developed in this [IRI] group has been control by workers who have taken power," those revolutionists would "give scope to the factors of consent and cooperation" in the development of their social plans. Faced with such a prospect, the technical vanguard thus had to choose whether to "face in the direction of the old economy which has given them their jobs, or whether they will go in the direction of their technical training." She clearly believed, with Veblen, that many of her like-minded comrades would follow the latter course. Because "a system which creates obstacles to the application of science to human society is neither a scientific nor a rational system," the course of thought and action appeared to be unarguable.[24]

Science held open to van Kleeck the possibility of individuals transcending self-interestedness with a collectively beneficial social logic. All that was needed was a "correct understanding of human relations and the health of the worker," although she assumed correctness to be a self-evident logical category and not a politicized judgment. Van Kleeck stumbled on the degree to which logic is neither universal nor unequivocal. At the same time, she persuasively claimed that "the community itself must discover its own program of action"; she grasped that goal setting was a social and not a technical process. Van Kleeck adopted a radical and proletarian politics as the process for the definition of those ends. Merely by comprehending the contentious nature of human society, she distanced herself from the settings in which she had established herself as a forceful advocate of managerial solutions to social problems. In doing so, she helped to expand the possibilities for the reconception of social rationality even as she failed to solve the conundrum of reconciling expertise with participation. Just recognizing that the conundrum existed distinguished van Kleeck from many of her scientistic peers, their greater fame and apparent accomplishment deceptive in their historical impact.[25]

ENGINEERS AND TAYLORITES

In opposition to van Kleeck's proletarian radicalism, an emphasis on mass production in the 1920s implied that the consumer society made possible by engineering and capital was an unquestioned benefit. Glenn Frank, the president of the University of Wisconsin, wrote in the *Magazine of Business* that "the masses have more to hope for from great engineers, great inventors, and great captains of industry than from the social reformers who woo them with their panaceas." In another article in the same series, Frank suggested that the introduction of the vacuum cleaner and the electric flatiron may have "meant more to the average woman—and been more prized by her—than the bringing of woman suffrage." The title of one article—"Back to the Spinning Wheel?"—clearly tied engineering to progress and made any protest of the machine age look silly. In contrast to the reform movements he derided, Frank represented the political dimension of applied science to be positive and uncontroversial, and in the process denied women a place in their own battle for full citizenship.[26]

By the time of the Depression, though, defensive tributes praised the formerly heroic engineer. The ASME, celebrating its fiftieth anniversary in 1930, commissioned the Yale dramatist George Pierce Baker to create an artistic celebration of engineering's triumphs. He responded flamboyantly by orchestrating "Control: A Pageant of Engineering Progress," presented at Stevens Institute of Technology in Hoboken, New Jersey. Featuring familiar historical technicians like Watt, Faraday, and Edison, "Control" also included allegorical figures like Mystery, Control, Conversion, and Imagination. The rousing finale featured symbolic characters, who, in the words of the ASME's most recent historian,

> return to the front of the stage. They chant a chorus of engineering
> accomplishments in a litany that emphasizes the words "power" and
> "beauty," while films are shown of automobiles, trains, steamships,
> and skyscrapers. Then, in a splendidly fantastic conclusion, and "in a
> great glow of light and color," Beauty, the last of the allegorical figures,
> emerges. She, it turns out, is the child of Control and Imagination, and
> it is she who has the last prophetic line: "I beckon ever into greater
> heights and flights." A grand march brings all the players back on stage,
> and in a fine Baker climax, President Hoover appears on the movie
> screen with a personal message for the occasion, fading out to "America
> the Beautiful" which the entire cast joins in singing, while "colored

lights flash and whirl over all, dying as the curtains slowly close and the pageant ends."[27]

The vision of the engineer conveyed by such bombast clearly comforted the faithful as skeptics continued to chip away at the profession. Ralph Flanders, a steadfast defender of engineering, addressed the critics forthrightly in *Taming Our Machines* (1931). While admitting that "the Machine becomes a personified evil for many," he held to applied scientific method as a source of social hope. "The first effective motive [for rebuilding society] is a *living faith*—primarily the faith of the engineer," he wrote. Turning crisis to opportunity, he continued: "We are invited to organize human society on the basis of human well-being, and the next step is the one of controlling the economic environment. Engineers have provided the tools and methods to make it physically possible." Flanders connected inquiry to action in explicit terms as the basis for such faith: "It is the duty of the scientist and engineer to analyze, and from the analysis to learn to manipulate and control— or at least adapt one's self to these natural forces." Flanders thus became one of the first to resume the search for an engineering vanguard to combat the Depression. In the commercial mania of the previous decade, the mythical engineer had faded from public political discourse as social scientists tried to rationalize social control. Meanwhile the businessman, often embodying managerial attributes, frequently replaced the engineer as an American hero.[28]

For the managerially inclined engineers who worked out of the Taylor Society, the 1920s offered apparent good fortune. Many of their theories and ideals underwent widespread adoption in business and industry, and the ASME set up a "management division" in 1920, in some measure superseding the society. The very acknowledgment Frederick Taylor had been denied in 1911 came almost a decade later, and by 1922 the new section was the largest of the society's professional divisions. When Herbert Hoover convinced the FAES to investigate waste in industry, Taylorites occupied eleven of the seventeen positions on the committee. C. E. Knoeppel, the former preparedness advocate, wrote a detailed questionnaire for the study, while L. P. Alford, Harrington Emerson, and Morris Cooke also joined the panel.[29]

Despite these promising opportunities, the Taylorites no longer carried themselves with revolutionary bravura. Industrial psychologists claimed new insights thought to render Taylorism less relevant. Salesmanship gained momentum and advertising increased in sophistication. Moderate conciliation between managers and labor unions replaced the bitter antagonism of the

Fig. 8-4. Allegorical figures from "Control," a pageant celebrating the fiftieth anniversary of the ASME. Left to right: Intelligence, Imagination, Mature Control, Conversion, and Finance. (S. C. Williams Library, Stevens Institute of Technology, Hoboken, N.J.)

sort encountered by early Taylorites at the Watertown Arsenal near Boston. Taylor would have been appalled by at least some of these tendencies. It is difficult to imagine his reaction to a 1920 article in *Industrial Management* as Warfield Webb called employee singing "a tonic that thrills and helps to build up a fading interest in work. The men go back to their benches with a newer feeling and interest in life." The managerial ideal so closely associated with engineers in the 1910s was a decade later more often held by businessmen and by social scientists like the industrial psychologist Elton Mayo.[30]

While such organizations as the SSRC, the NBER, and the Brookings Institution obtained significant amounts of money—$21 million from the LSRM alone between 1924 and 1928—to find the social facts from which scientific reform could proceed, the Taylor Society steadily lost membership, influence, and, eventually, its name. Why did this fate befall the group that pioneered the application of science to the correction of imperfect social relations? In these years after the war, the reform-minded managers and engineers in the Taylor Society had competing claims on their time. Many withdrew to devote their energies to their firms or communities. Others died, and few new members joined, even though "junior members" were recruited in schools of business administration. Some, including leaders of the movement such as Harlow Person and Henry Dennison, retained their Taylor affiliation while also working with social scientists on other projects. The figures tell only part of the story:

1923	1924	1925	1926	1927	1928
863	791	763	706	688	639

Taylor Society membership in the mid-1920s was deceptively high in that an increasing number of the members were foreign and therefore less likely to contribute to the vitality of the group in the United States. The Taylorites did not require a Gantt chart to know which way the wind blew.[31]

By the end of the decade, caught between their internal problems and such vigorous competitors as the American Management Association, the leaders of the society decided to pursue the same apparently simple strategy that had enriched the social science research ventures. The successful enterprises devoted to improving social technique all had significant financial support from the newly important foundations, so the Taylor Society sought to prove its worth to the foundations by arguing it could be home to efforts to investigate and improve all aspects of American life. In 1928, amid this campaign for primacy, the *Bulletin of the Taylor Society* asked its members the same question it was putting to the philanthropists: "Can the planning which has proved to be effective in coordinating the departments of the individual enterprise be established on a plane on which enterprises are but departments of one integral social enterprise?" Putting on a multitude of efforts at social rationalization cost money, and Harlow Person, the managing director of the society, wrote to Morris Cooke, his still-interested predecessor, of his strategy in 1925. "I am to do my best," he told Cooke, "to interest Fosdick and that bunch" in helping to insure the survival of the organization.[32]

Raymond Fosdick and, more crucially, Beardsley Ruml refused to include the Taylor Society with other Rockefeller efforts to apply scientific social inquiry to America. In mid-April 1927, Person repeated the message to Cooke: "We must pitch in and dig up funds." A "Conference Dinner," held in New York on April 28 of that year, represented a significant attempt to return the

Taylor Society to the center of scientific social reform. The theme of the dinner, according to the affair's seating chart, was premised on "the common interest" of the participants: "The maintenance of an adequate standard of living for the whole community in all countries." Clearly a corporate welfare state—or world—had supplanted the firm as the group's focus. Taylorites devoted the dinner to asking how their society could "serve more effectively . . . as a common center for the activities of business executives, engineers, leaders of labor, publicists and social scientists."[33]

The Taylor Society directors invited many leaders of the movement to apply science to social problems: foundation officials, social investigators, journalists, and business executives. The rosters of these groups show how Person and the Taylor leadership understood the state of social engineering.

The foundation officials who were invited but did not attend included Evans Clark, the labor editor of the *New York Times* who eventually became director of the Twentieth Century Fund and was also married to a noted liberal, Freda Kirchwey of the *Nation*; Edmund E. Day; Frederic Delano, a board member of the Institute of Politics (which in 1928 became the Brookings Institution) and later on the NRPB with Wesley Mitchell and Charles Merriam; Edward Filene, a Boston department store executive and founder of the Twentieth Century Fund in 1919; Raymond Fosdick; Frederick W. Keppel, the president of the Carnegie Corporation who later declined a grant proposal made by the Taylor Society in 1933;[34] and Beardsley Ruml.

The foundation officials who attended were John M. Glenn, the president of the Russell Sage Foundation; and Mary van Kleeck.

Several social investigators were invited but did not attend: Irving Fisher, the Yale economist previously active in the efficiency crusade while working on currency reform, eugenics, and other social solutions; Edwin Gay; and Edward Eyre Hunt, the aide to Herbert Hoover who edited the volume *Taylorism after Taylor* and was on the Taylor Society payroll in 1927 while on assignment.

Social investigators in attendance included Paul U. Kellogg, the editor of the *Survey* who, beginning with the Pittsburgh survey, bridged social work and reform journalism to apply scientific findings to society; Wesley Mitchell; and Joseph Willits, dean of the Wharton School who had been a Taylor Society member in 1916 and later became a board member of the Rockefeller Foundation.

Journalists Herbert Croly of the *New Republic* and Walter Lippmann of the *New York World* were invited but did not attend.

One other invited guest deserves mention: Hugh Frayne of the American Federation of Labor. Because of a history of intense antagonism between early Taylorites and labor, the organizers of the dinner impressed upon the foundations they courted the close cooperation between labor and capital within contemporary Taylorism. Frayne was well known to the readers of the *Bulletin of the Taylor Society*, but the records left his absence at the dinner unexplained.

Being snubbed by the seven philanthropic managers as well as by Croly, Lippmann, and Frayne must have disheartened the directors of the society. Day and Ruml favored men like Mitchell to lead their investigations, while Keppel had little time for managerial solutions, preferring to work less programmatically on social improvement. Glenn and van Kleeck had limited resources with which to address more concrete social needs despite their sympathies. One small grant from the Twentieth Century Fund supported a membership drive and expanded publicity, but this level of support fell far short of what the social scientists were receiving and what Person and Cooke sought. Day, Gay, and Dennison did help form the Business Research Council, a group intended to parallel the SSRC, but the council drew little support from competitive business executives who preferred internal rather than centralized research.[35]

At the dinner itself, the Taylorites shamelessly promoted their attempt to unite management and governmental reform. Cooke asserted that "if we teach the average workingman to detect a faker in industry, he will make the application later in the governmental field." The longtime Taylorite engineer Robert Wolf centered his appeal on the promise of labor peace: "We in this country have an opportunity to keep the labor movement out of politics by giving it a chance to function legitimately in industry." Repeating the basic message for those assembled, Person claimed that the society could be a home for many investigators and reformers. The initial meeting of the Society to Promote the Science of Management in 1912 began "the development of an institution in which executives, engineers, economists, psychologists and others interested in the social sciences are now working together on problems which are of common interest." The modern Taylor Society, he promised, "is a society which can without much difficulty tap the facilities of all those groups." The call for integration echoed the founding rationale behind the SSRC, which by 1927 had become, with Rockefeller support, a powerful entity in the field.[36]

Van Kleeck made an even stronger argument for management engineering and for a theory of social change based in engineering logic. After noting the *Encyclopaedia of the Social Sciences* and other developments in the funding of social inquiry, she predicted that "many of those studies . . . are going to reveal dislocations in the economic organization, which again goes back to the workshop." She then laid out an ambitious vision of a coherent society:

> If there is to be any really sound regional planning there must be not merely an architect to design buildings, but the economist must inquire, What is this region for? And the social worker must say, Where are the people going to live, and where are they going to be educated, and where is their community life to be organized? The political scientist must inquire, What relation has government to this scheme of things? The engineer must ask, How can we plan for people to get to their homes and to their work? There must be some orderly arrangement.[37]

Even after appealing to logic and common sense, the Taylor Society won neither grants nor the opportunity to coordinate the movement toward social rationalization. The Rockefeller Foundation, the agency most likely to back Person, continued to focus its attention on established academic and institutional ventures in social investigation throughout the 1930s. Because the Rockefeller Foundation discards unsuccessful grant applications after a few years, and because foundation support had become a prerequisite for credibility, the Taylor dinner affords the historian a rare chance to see one of the roads not taken toward rationalized social reform.

. . . .

These three attempts to implement and popularize social engineering precepts shared a commitment to scientism even as they had sharply differentiated constituencies: the fair-going public for Odum, technicians and workers for van Kleeck, and management engineers for Person and Cooke. The engineering method provided the intellectual leverage that each of these reformers tried to energize—after they got a foundation suitably interested. All considered themselves social engineers even though they approached social problems from different angles. Odum sought public appreciation for and acceptance of social scientific work in universities and institutes; van Kleeck envisioned an international worker-technician movement employing logic to enable consensus; and Cooke and Person looked to blur the distinction between corporate and public aspects of welfare, education, and administration. That all three programs were considered scientific by their proponents, despite their divergence of outlook, suggests that science was not as unambiguous as they had imagined. No matter how powerful their ideas or ironclad their reasoning, these advocates of rational reform failed. Their mobilization of science was undone by politics and by power of a more traditional sort, exercised by the philanthropic foundations operating in the twilight between public and private definitions.

9

SOCIAL ENGINEERING IN THE DEPRESSION

I: OUTSIDE THE NEW DEAL

During the Great Depression, unemployment, home-lessness, and other dislocations disturbed American citizens, leading some to question the very premises of industrial capitalism. Public interest in rational solutions surged once again after a decade of behind-the-scenes efforts within the Hoover, philanthropic, and social scientific circles. The crisis brought proposals from all camps—camps as diverse as free-market libertarians, mainstream liberals, and avowed Communists. The social engineering model, therefore, enjoyed only partial, but growing, support after 1930.

Although a highly permeable membrane often divided government from other institutional centers of expertise, a line can be drawn between the extra-governmental planning advocates and the New Deal projects that drew most heavily on the rational reform tradition. The theory of cultural lag animated a wide variety of social commentators who saw dislocation between technical competence and political abomination. "The mechanical inventor has given us Chicago, and the lack of the social inventor has given Chicago its city government," wrote one disgusted observer. "The mechanical inventors give us bombing planes, while Cro-Magnon politicians still chip flints." By 1930, engineering had diffused in a number of directions, finding expression in managerialism, journalism, social science, and alternative political organizations.[1]

ASSORTED TECHNOCRATIC IMPULSES

At the Rockefeller Foundation division of social science, social planning based on scientific investigation continued to be a topic of discussion and investment. Under Edmund E. Day's ongoing leadership, research projects became even more closely linked with social improvement as the Depression threatened the stability of institutions throughout the world. Accordingly, in 1933 the trustees initiated a program of economic stabilization and community planning. Under the new plan, "the social sciences . . . will concern themselves with the rationalization of social control." The concentration on recruitment of qualified investigators and on institutions such as the NBER remained strong. Day's justification of this approach continued to draw on science as the most efficacious social method. If "the afflictions of modern competitive society" are to be adequately addressed, he argued in 1931, "a high order of technical skill and social intelligence must be brought to bear." America needed "positive and vigorous development of social intelligence—the understanding and control of social institutions and social processes in the solution of pressing social problems." The rationale had changed little from nearly a decade before. Investigators using scientific modes of thoroughness and objectivity would generate facts that efficient reformers could apply to social problems, and effective social solutions would in turn validate the investigations.[2]

Such continuities also coexisted, however, with a severely depressed stock market, with a rapidly darkening world picture, and with continuing Rockefeller Foundation efforts to address international problems. In the 1930s the board scaled back the heroic scope that characterized Ruml's and Day's projects of the 1920s, and the degree to which engineering validated the program also decreased over the years. A conscious shift in direction in 1934 markedly diminished support for university research as more focused—and more accountable—projects received closer consideration. The SSRC and the NBER continued to win Rockefeller funding, but ambitious projects such as construction of research buildings at Chicago and Yale, the *Encyclopaedia of the Social Sciences*, and *Recent Social Trends* were no longer underwritten, casualties of doubts about importing natural scientific thinking into social affairs.

. . . .

In the same period, journalists helped to widen the audience for programs of rational reform after a period of quietude. Two men in particular—Stuart

Chase (1888–1985), the author of many books and articles on industry and waste, and George Soule (1887–1970), an editor of the *New Republic*—entered the Depression as apostles of scientific approaches to social problems and then intensified their calls. The two men cofounded the Labor Bureau in 1920, and its survey of waste updated the efficiency ideal of the 1910s into a more quantitative, "factual," variant. Their other work did so as well.

A college classmate of T. S. Eliot and Walter Lippmann, Chase shared with the technocrat Howard Scott a lasting debt to Veblen's *Engineers and the Price System*. Chase, a commentator noted, "has given us Veblen, Veblen, Veblen, nothing more—and a little less." After World War I, Chase began to reveal his characteristic scientism, behaviorism, and hardheadedness. In his desire to "grow ever more radical by getting ever nearer the roots," he wanted to "take Wall Street to pieces the way Jacques Loeb [one of Veblen's influences] took a starfish." To argue for expert management of material abundance, Chase single-mindedly developed a political economy premised on mass-produced plenty. Unlike many social engineering advocates, he actively considered readjusting consumption as well as production and founded the Consumers' Union to bring rational methods into the economy from the demand side. By 1935, Chase had balanced material security against the ballot and found the latter wanting. "I would exchange all the political democracy ever heard of, and all the constitutions, and all the founding fathers," he declared, "for the real democracy of the universal right to be born clean, to grow strong, and not to be crawling on one's belly to a petty tyrant for a job. I would suffer an economic dictatorship to secure this happy state." Such extreme statements merely culminated arguments present since the early 1920s.[3]

In 1922 Chase had noted that "an engineer has coined the word 'technocracy,' which speaks for itself," and thereafter he continued to rely on engineering for America's decisive political regeneration. For him, the job of a radical was "the remorseless pursuit of what the psychologists call problem-solving thinking—as distinct from rationalizing, reverie, and drifting with the tide." While he envisioned "engineers of the humanities," the future lay with professional engineers who could remake the material world. "Mr. Engineer," Chase chided in 1931, "you have played the shrinking violet long enough. . . . Plato once called for philosopher kings. To-day the greatest need in all the bewildered world is for philosopher engineers." This breed of administrator would need to "know statistically all about everything," an omnipotence that resulted from the technician's "altogether realistic perception of cause and effect." Politics, as it had for many technocratic reformers, appears in Chase's work as the evil to be overcome: "Neither mysticism, political rhetoric, nor contemplation of the navel will get kilowatts out of Niagara." Like Gantt, Veblen, and Cooke, however, Chase stood on the platform waiting for a locomotive of social change that failed to run on his timetable.[4]

Soule, meanwhile, had originally joined the *New Republic* in late 1914, with Alvin Johnson and Randolph Bourne. After leaving the journal during the war, cofounding the Labor Bureau, and being named a director of the NBER in 1922, he returned to the *New Republic* in 1924. His political writing focused on economics, especially productivity. Less given to hyperbole than Chase, Soule nevertheless endorsed the ability of technicians to reshape outmoded conceptions of administration and governance. *A New Deal* (1933) and *The Coming American Revolution* (1934) outlined his program for reconstructing the damaged economy.

Unlike many like-minded writers, Soule continually refrained from offering apocalyptic visions of overnight transformation; he always predicted a period of confusion and dislocation preceding "the final disappearance of government by private profit-makers over the means of production, [and] a chance for social management to learn its task by experience." The centralized decision making implied by such a model drew on the application of science. Because technology had developed sufficiently, dreams of "synthesis, coordination, rational control" no longer could be dismissed as utopian. In fact, in keeping with the theory of cultural lag, it was humanity's duty to apply technical insights to social groups, to consciously adjust America to technological innovation. Public opinion would then follow the visionaries: "As more and more people—both engineers and others—come to understand the *inherent superiority of the engineering approach*, the traditional business way of doing things is bound to lose its popularity."[5]

Two other themes connect Soule to the broader discussion of rational reform. While not endorsing the Soviet system wholeheartedly, he did allow that the technical aspects of its experiment with economic and engineering management were "not wholly foreign" to the American situation. He also foresaw spiritual rebirth as an outgrowth of technological development. As he put it, "Instead of being baffled and burdened by an irrelevant environment of social forces," humans could master their lives and have "a warm and active bond with our fellows."[6]

Such journalism helped to move the rational reform debate out of the conference rooms of philanthropies, the Commerce and the Agriculture departments, and social scientific organizations into the agora. By the time this movement was taking place, however, the pace and confusion of American politics prevented many readers from concentrating on this particular variation on the planning theme. Considering that a bibliography of such texts listed over 1,500 titles in 1933, Soule and Chase had become to some extent merely voices in the crowd.[7]

. . . .

While many proposals for social reconstruction relied on engineering only in superficial or symbolic ways, a small but important group of managers drew on their experience in industrial firms or on insights from the managerial reform movement as they called for "planning." Harlow Person, E. A. Filene, and Henry Dennison, in particular, show just how thoroughgoing the emphasis on industrial productivity as a political and economic elixir had become in the early Depression. These veterans of Taylorite reconstruction of the firm watched younger, more innovative, or radical reformers build on their industrial ideology and, in the process, transform it.

Henry Dennison (1877–1952), for example, moved easily from the presidency of the Taylor Society to other attempts to streamline social relations. His emphasis on abundance made him unexceptional in this period, especially insofar as he endorsed a modified Hooverian associationalism that lost considerable credibility between 1929 and 1933. This outlook put him in close agreement with the corporatist plan put forth by General Electric chairman Gerard Swope, and the two in fact corresponded after discovering their affinity. Nevertheless, Dennison retained a place for academic social science in his thinking and urged political scientists to see society the way students of industrial management did. Social research, he argued, had to move "from the descriptive to the analytical, and from the analytical immediately to the engineering point of view." Such a perspective belonged not to "the historian or the moralist, but [to] the student of applied science, the engineer." Like dozens of others, Dennison insisted that advances in physical science mandated the use of similar methods in social reform; "progress in physical engineering will continue and so make necessary a still more rapid progress in social engineering." He eventually joined the New Deal in an official planning capacity, where he got a chance to apply his theories.[8]

The programs of many social engineers tended to focus on the national level, but one important sector of the planning movement deserves more extensive attention than it can receive here. Howard Odum and Lewis Mumford, among others, lent leadership to regionalists, who attempted to balance the distinctiveness of geographic units with the gains attending to foresight, economies of scale, and coordinated programs of action. An organicism arising from a fear of urban homogenization attenuated the bureaucratic tendencies of larger-scale plans. Odum, for example, saw regionalism as a way to retain the South's sense of place while transcending the racial backwardness he decried. Mumford's humanistic criticism revealed itself most sharply in the regional movement as he steered clear of the autocratic pomposity and giant abstractions that fascinate yet ultimately repel in *Technics and Civilization* and elsewhere. On the whole, engineering sensibilities coexisted much more successfully with human communities in regional plans than in national ones.[9]

One final manifestation of the planning impulse fits into the puzzle here: the National Economic and Social Planning Association. This group once again looked to systematic consideration of human needs and resources as the key to providing the American people with "the highest possible material and cultural standard of living." One NESPA member called planning a "great unifying principle for communities which have been torn by economic and industrial strife," recalling the same promise made for efficiency two decades earlier. The term was defined extremely loosely, including the regional and city, national and international arenas. Consequently the articles in the group's journal, *Plan Age*, reflected both a broad constituency and a severe lack of agreement on basic principles inevitable in such a vague notion. The planners saw insufficient technical capacity rather than philosophical justification to be the primary impediment between crisis and abundance, so many of planning's virtues were asserted rather than proved.[10]

In the epitaph for the association written in late 1940, Soule noted the origins of the group's "apparatus and methods" in scientific management circles. But these planners expanded the scale of rational intervention from the business enterprise to the nation and world. Significantly, the group reorganized in 1941 to address the postwar situation, and it was soon joined by countless other inquiries, proposals, and committees on the same topic. By 1940, however, partially because of what happened to the NRA, rational and centralized control of an economic unit as vast as the United States had proven impossible. The engineering model, with its efficient hierarchies premised on adequate knowledge and common assumptions, no longer served as the unchallenged referent for modernized reform and reconstruction.[11]

. . . .

The Technocracy movement of the 1930s, in some senses social engineering's culmination, began with three men whose system of "energy values" sought to evaluate the economy in terms more relevant than profit and loss. Howard Scott's slogan—"government by science; social control through the power of technique"—encapsulated the hopes of a political impulse that eventually claimed a geographically far-flung and sociologically diverse body of supporters. Originating in the works of Veblen, Taylor, Cooke, and Gantt, Technocracy reformulated the dream of social engineering into a bold and ultimately self-contradictory program. Was technique to be the servant of newly defined human aspiration, or was it to supplant the imprecise language of values with an unimpeachable logic of productivity? Scott, Harold Loeb, and Walter Rautenstrauch, the leading triumvirate of the movement, could not agree, so no national figurehead could unite Scott's New York Technocracy, Inc.; Rautenstrauch's Committee on Technocracy; the All American Technological Society of Chicago; and Loeb's Continental Committee on Technocracy.[12]

For a moment, some technocrats had their chance. Scott drew attention to the movement by writing an article for *Harper's* magazine but thereupon underwent journalistic scrutiny that exposed him as an engineer without credential. A national radio audience then listened as Scott addressed the New York Society of Arts and Sciences in January 1933. The previously flamboyant impersonator bombed, losing his temper and refusing to answer several questions. Disowning the performance, the technocrat press later claimed Scott had been drugged, and Loeb split with Scott over the degree to which the latter idolized engineers. Loeb then undertook a thorough investigation of American productive capacity, finding that "the resources, man-power, equipment, and technology existing in the nation are ample to provide a high standard of living for every inhabitant of the United States." He did stop short of insisting that only technocracy could make that abundance possible. The masses who joined and organized the various groups expressed a conviction that "electoral politics seemed part of another world, a kind of fantasy land that had nothing to do with the day-to-day reality of work and consumption," in the words of the historian of the movement. Many unresolved discontinuities between traditional politics and the lives of discontented citizens forced some of them to confront the concept of democracy with a production-based vision of the state.[13]

Like Chase focusing on the potential abundance left unattained, Technocracy Inc. promised material plenty as the reward for organization, along industrial and functional lines, into Technates, transnational units of technological control. Technocracy, the organization's *Study Course*[14] claimed, would not "destroy the Price System. The Price System destroys itself." This focus on political economy as only a supporting infrastructure for manufacturing and industry merely manifested a deeper precept: technocracy, a political movement, denied that politics existed. The "competent functional organization" proposed by the technocrats *has no political precedents. It is neither democratic, autocratic, nor dictatorial.*" Science justified this escape from politics: "There is only one science, and there is no essential difference between science and engineering. The stoking of a bunsen burner, the stoking of a boiler, the stoking of the people of a nation, are all one problem." This regimented program was the most extreme rational reform system put forth, but it differed from the rest in intensity, not in kind. Abundance at any social price, science as a trump card over all opposition, and a denial that politics even existed—all of these elements could be found in the plans that drew on engineering for their validation.[15]

BEARD AND DEWEY

The careers of Charles Beard and John Dewey extended far beyond the Depression years, but these men are considered here for several reasons. Their interdisciplinary and eclectic interests fell increasingly out of fashion as intellectual and professional specialization, exemplified in the SSRC social scientists, took precedence. By 1930 both Beard and Dewey marked the end of an era, for no younger minds came forth to carry on their synthetic legacies. Pragmatism, too, informally bound the two men, as Beard looked to performance as a standard of validity while refusing, like Morris Cohen, to wear the label. Finally, both men understood the nation and the world to be in sufficient trouble that each entered the realm of practical politics and wrote on matters of everyday interest in mass journals. Both, finally, faced a conundrum: they wanted to capitalize on the success of applied natural science and to delineate a stable, modern value system to buttress and guide the power of scientific rational technique.

A democratic scientism that originated in the mid-1910s persisted in the work of Charles Beard (1874–1948), but by 1930 he struggled to reconcile his faith in scientific method with the realization of its potential for technocratic abuse. A materialist view of history initially led him to downplay ideological forces as he cut to what he understood to be the gristle of the human past: political and economic conflict. Money and power, tangible forces both, merited attention as causal levers. In the reform realm, the actors most capable of controlling those forces were experts. Beard wrote Raymond Fosdick

Fig. 9-1. Charles A. Beard (DePauw University Archives and Special Collections, Greencastle, Ind.)

in 1922 that "the sword won't do the job any more." Instead, he continued, "the social engineer is the fellow. The old talk about sovereignty, rights of man, dictatorship of the proletariat, triumphant democracy and the like is pure bunk. It will not run trains or weave cloth or hold society together." In 1925, the sentiments remained the same. Even more than in the 1910s, Beard regarded engineering as a political resource. Casting off "all my lingering suspicions about the value of science," he was "more convinced than ever that we shall make progress by applying the methods of natural science to the study of government and administration."[16]

Beard's apparent confidence soon gave way to a more questioning attitude as challenges to his materialistic, scientific reformism came from several sources. The model of science as the acme of certainty had to be revised in the light of institutionalized uncertainty in physics and mathematics. At some point Beard read Ernest Hobson's *The Domain of Natural Science*, which asserted that unbridgeable discontinuities precluded an analogy between natural science and programs of social reform. In addition, a materialist interpretation of history, with its insistent exclusion of the "soul stuff" Arthur Bentley so detested, did not square with the European social theories Beard was encountering at this time. The work of Benedetto Croce, Hans Vaihinger, Karl Heussi, and later Karl Mannheim and Friedrich Meineke forced Beard to admit that economic and material forces could not by themselves account for the past any more than could abstracted ideas. Preparing *The Rise of American Civilization*, it appears, forced Beard to allow for the role of ideas in the formation of something so inclusive as the concept of civilization the way he and his wife, Mary, employed it. After the publication of *The Rise* in 1927, Beard began to speak out more vocally on the limitations and deceptions of scientism in historical and political scholarship.[17]

Combative on many fronts, Beard did remain committed to certain aspects of the engineering analogy and attacked what he understood to be the inadequacies of modern inquiry. Repeatedly invoking classical figures—Aristotle, Machiavelli, the authors of *The Federalist*—as the antitheses of modern academic scholars, Beard lashed out at the pretensions of political science. In an article on the discipline in 1929, he first softened his call for a science of reform: "Without doubt the scientific method is highly useful in political affairs, but it has decided limitations. Both logic and statistics can be bent to serve many causes." A page later, he echoed Woodrow Wilson's critique of twenty years earlier, repeating the assertion made by Ernest Hobson. "No science of politics is possible; or if possible, desirable. . . . The method of natural science is applicable only to a very limited degree and, in its pure form, not at all to any fateful issues of politics." Purity of method could not substitute for moral conceptions of the political good. Beard told the APSA that science was incapable of telling "anyone what to do in any large human situation, what is most valuable, what is worth doing. . . . Without ethics, political science can have no more vital connection with life than have the tables of an adding machine." Linking virtue and politics would become even more crucial for him as the Depression wore on.[18]

In 1929 and 1930 Beard gave considerable attention to issues of industrial technique in the democratic state. He edited or co-wrote three major volumes, all of which dealt with this theme: *The American Leviathan: The Republic in the Machine Age*; *Whither Mankind*, a symposium on the fate of humanistic values under industrial capitalism; and *Toward Civilization*, a volume of rebuttal by prominent (and thoughtful) engineers. The high caliber of the contributions distinguishes the latter two volumes; among those writing essays were John Dewey, Lillian Gilbreth, Lewis Mumford, Bertrand Russell, Elmer Sperry, and Sidney and Beatrice Webb. Beard's introductions and conclusions, careful and articulate, indicate his own vacillations and misgivings. With *Leviathan*, these short writings portray the dialogue of internal reconsideration.[19]

Beard began *Whither Mankind*, the volume on the humanities, by asking whether "the imagination of an Einstein, a Bohr, or a Millikan may [not] well transcend that of a Milton or a Virgil. Who is to decide?" After sixteen essays on the decline of humanistic values, Beard concluded the volume by refusing to assent to wholesale condemnation of the machine. While calling for ethical awareness from political scientists, he asserted that the common theme in the collection was "that by understanding more clearly the processes of science and the machine mankind may subject the scattered and perplexing things of this world to a more ordered dominion of the spirit. This is the paradox of the symposium." For all of its flaws, the machine—infallibly logical and superseding mere humanism—remained the source of hopefulness, for Beard more than for most of the contributors.[20]

In *Toward Civilization*, the volume by the technologists, Beard showed more brio; he and Mary had recently contradicted Oswald Spengler by concluding *The Rise of American Civilization* with the declaration that it was "the dawn, not the dusk, of the gods." The engineer had done great things and was poised to take a yet more fateful step. "Heir of the past, path-breaker in the present, the engineer, by virtue of his labors, is in a strategic position" to assess the civilization which he has created. Engineers of the future, predicted Beard, "will give increasing attention to the values inherent or implied in their work." In his contribution to the volume (which Beard liked "immensely"), Ralph Flanders reinforced the prestige of engineers but explained why their method would be ill suited to politics. "Engineering," declared Flanders, "may be defined as science which works." To maintain his reputation, the engineer "refuses to deal with anything unless he has proved that it will work." Such a refusal would apparently banish politics from the engineer's kingdom.[21]

Nevertheless, inspired by the high quality of the essays in the volume and dismayed by the prospects of the Depression, Beard contradicted Flanders's characterization. He envisioned a rational public not unlike that invoked by Hoover, and claimed that a greater infusion of the spirit of engineering into politics would diminish "the necessity for coercion," in contrast to his own past theories of power and conflict. Confident of the eventual and voluntary adoption of humanistic principles by technicians, he closed the essay on a soaring note, asking for "more engineering, not less, engineering informed with respect to its human implications, controlling unlimited power, mastering the nature of materials, adapting them to mankind and mankind to them, conscious rationality triumphant, not as purpose only but also as an instrument worthy . . . of 'conquerors by the grace of God.'" Such reasoning could not persuade all of his readers. The reviewer Howard Mumford Jones wondered why the engineers tended to live, judging from their essays, "in a world without politics, without social problems, without an intricate economic system." As a result, Jones called the "social and economic thinking" in the book "hopelessly naive."[22]

In *The American Leviathan*, Beard tried to fuse democratic aspirations with an overwhelming faith in technique, his criticisms of political science bracketed momentarily. Acknowledging that intuition and ethics could not be excluded from government, he reiterated that "science and machinery do not displace cultural considerations"; civilizations had to combine "the noblest philosophy with the most efficient use of all the instrumentalities of the modern age." Science remained central, for at the core of a great society pulsed "the spirit of engineering—the spirit of law as distinguished from chance." Once again, "mankind's conquest of the future . . . will largely depend on the successful application of the scientific method to the affairs of government as well as private enterprise." For all of this technocratic bell ringing, Beard still attempted to dream the dream of scientifically empowered masses using the methods and capabilities of the machine to remake the world. "History-making in the machine age" should involve the masses rather than merely "a small aristocracy of conquest." By 1930, however, he

Fig. 9-2. Before electronic and digital computation, artificial intelligence was mechanical. The "Brass Brain" provided one impressive example. (Charles Beard and William Beard, The American Leviathan: The Republic in the Machine Age *[New York: Macmillan, 1930])*

gave this claim less and less emphasis. It was the vision of technique, albeit technique recolored by human values, on which Beard ultimately placed his bets.[23]

During 1931 Beard put forward his version of a five-year plan to join many others. In it he attempted to balance, once again, rational technique with human values. The "coming philosophy of ethical reconciliation," he wrote, would not "take aboard any of the epic theologies" but instead would center on the "good life." It would also "rise above parties, sects, and mass producers" and thus would implicitly reject some aspects of consumer society. Nevertheless, the new and good life would "be planful, because the good life cannot be lived without scheme and control, and the supreme instrumentality of our age, engineering, is planful in operation." Beard's five-year proposal, however, placed substantial confidence in bureaucrats, not engineers: councils, boards, and syndicates abound in the nine-point plan. The conflict between efficiency and democracy that he had been exploring for decades remained troublesome.[24]

By the early 1930s Beard had also undertaken a significant scholarly reformulation in addition to his oft-articulated social pronouncements. Still fascinated with the explanatory power and apparent neatness of the natural sciences, he was stymied by big causal questions. In *The Rise of American Civilization*, for example, he wondered whether the fact that "political democracy and natural science rose and flourished together" implied any "deep connections in their inception." Like Henry Adams and, to a lesser extent, Thorstein Veblen, Beard searched for some inclusive science of history that could meet minimal criteria for interpretive adequacy and still not threaten humanity with preordained or imprisoning determinism. Tumultuous world events after 1928 forced him to reformulate both his historicism and his epistemology. In a number of essays, most notably "Written History as an Act of Faith" (1934), Beard confronted the split between fact and value, the myth of historical objectivity, and the impossibility of definitive criteria for anything. For both the writing and living of human history, he needed to reunite empiricism and aspiration.[25]

At the outset of the technocracy craze, Beard warned against undue reliance on a conception of the engineer as a social savior but soon found himself in close agreement with some technocrats. He wrote the introduction to Graham Laing's *Towards Technocracy*, equating the technocrats with Madison, Jefferson, and Adams and noting the parallels between crises. "It was evident to [the founders of the American Republic] that the political and economic machinery had broken down and was not functioning with an [adequate] efficiency," Beard wrote. He later praised a technocrat who stepped forward to "utilize science to discover the conditions, limitations, inventions, and methods involved in realization" of social dreams, as an earlier review had hoped.[26]

In his review of Harold Loeb's *Chart of Plenty*, Beard waxed ecstatic: Loeb's survey of social and industrial resources, "the most important book of the twentieth century that has come within my ken," was "the first attempt . . . to apply the rationality of engineering . . . to the central problem of American life and economy." For all his statistical inaccuracies, Loeb had bolted the first girders in the skyscraper of the new order: applying engineering rationality to social problems "is the only approach that promises a way out of the present defeatism and social degradation. When and if economists, politicians, statesmen, labor leaders and feminists get around to seeing it, we may expect the fog to lift and something be started that will astonish mankind."[27]

Such optimism, voiced in early 1935, marked a return to the same hopefulness with which he had concluded the 1927 edition of *The Rise of American Civilization*, when the Beards chided Spengler for his pessimism. In 1933 they updated *The Rise* to include a section on the Depression and revised the original conclusion about their time being the dawn, and not the dusk, of the gods. In its place was a much darker, pessimistic motif, reflecting Beard's frustration at being caught between materialism and idealism, the past and the future, the actual and the possible. The reference could also be autobiographical; Beard had lived nearly sixty years, many of them spent as a social conscience. A vivid image portrays the intellectual reformer as Sisyphus: "So Thought, weary Titan, continued to climb as for two thousand years the rugged crags between Ideology and Utopia." To make the ideal real, Beard looked to science, but his decision gave him neither rest nor certitude.[28]

. . . .

The stature of John Dewey (1859–1952) as an intellectual mentor to many social scientists, great as it was, may be exceeded only by the degree of misunderstanding his writing and presence engendered. Controversy that began with Randolph Bourne's disillusioned "Twilight of Idols" in 1917 has endured. Given the wide and profound influence Dewey exercised over social scientists, educators, and other reformers, he could be, and was, read in a variety of ways in order to validate a range of stances. As it is impossible and undesirable to establish fixed interpretations for the whole of his corpus, the point here is to explore the polarities that Dewey sought to fuse: self and society, democracy and logic, and means and ends. Paralleling Veblen, Dewey became a sometimes unwilling herald for intellectuals trying to justify social engineering in the early twentieth century.[29]

Dewey's social thought must be considered in the context of his political commitments in the 1930s, especially his work toward a third political party and his contributions to the journal *Common Sense*. That periodical, under the editorial direction of Alfred Bingham and Selden Rodman, provided Dewey with a broader forum for his political activities when it merged with the League for Independent Political Action in 1933. The party's slogan—"Make machines your servants! Outlaw poverty!"—resonated with those of producerite revisionists across the political spectrum, from Soule and Chase to Scott and Loeb. In the meantime, the league suffered from divisions between socialists and nonsocialists, and these camps only divided further in defiance of social engineers' faith in scientific method as a guarantee of consensus. The league eventually merged with the Farmer-Labor Party Federation in an attempt to capitalize on third-party success in the upper Midwest, and the group changed identity yet again in 1935 when the American Commonwealth Political Federation was formed in Chicago. When Franklin Roosevelt won his landslide reelection in 1936, Dewey retreated from electoral politics, but his experiences, in true pragmatic fashion, gave him additional insight into the revitalization of American politics.[30]

In its name and editorial positions, *Common Sense* stood for tangible, rational politics in opposition to the vested interests that restricted production, employment, and consumption. The magazine invoked Dewey when it described its "creed" in 1937: "That intelligence and good will can be applied to ordering social institutions, in order to achieve planned abundance; and that this change can bring a vast extension rather than restriction of human freedom." Among the contributors were many promoters of various social science and social engineering ideas: Thurman Arnold, Stuart Chase, Harold Loeb, Lewis Mumford, Harlow Person, Howard Scott, George Soule, Mary van Kleeck, and Henry Wallace. From the world of arts and letters, James Agee, W. H. Auden, James Baldwin, Thomas Hart Benton, Theodore Dreiser, S. I. Hayakawa, Langston Hughes, Thomas Mann, Diego Rivera, Delmore Schwartz, and Edmund Wilson also appeared. Daniel Bell worked there for a time. One index of the magazine's audience is provided by an advertisement for subscriptions to the *New Republic*, which offered respondents a complimentary copy of Veblen's *The Engineers and the Price System*.

*Fig. 9-3. John Dewey (John Dewey Collection, Special Collections, Morris
Library, Southern Illinois University, Carbondale)*

In its enthusiasm for planning, *Common Sense* endorsed engineering as one of several resources for rational reform.[31]

Dewey's enduring resistance to dualisms—self and society, science and morality, theory and practice—apparently resulted from his partial rejection of the Hegelian idealism of his early years. Decisive answers could never result from the public consideration of ideas, for the code of scientific investigation that comprised the substrate of Dewey's pragmatism implied that the "rightness" of ideas could not be established through a hermeneutic process: "The idea that the conflict of parties will, by means of public discussion, bring out necessary public truths is a kind of political watered-down version of the Hegelian dialectic, with its synthesis arrived at by a union of antithetical conceptions." In some ways, Dewey squandered his intellectual inheritance from Charles Sanders Peirce, never adequately appreciating signification. He posited that instead of relying on symbols, the usual tokens of political debate, social intelligence needed collectively to plan and design. But the issues of logic, participation, and scientific practice grow troublesome. In the flight from duality toward unity, Dewey may have overlooked a conflictual dimension in politics that proved less than amenable to scientific determination. Moving from idealism to instrumentalism, he dismissed illogical expressions that, while failing tests of empirical verification, often resonate with human experience.[32]

Dewey's emphasis upon the collective nature of human existence, a widely understood cornerstone of the pragmatic outlook, requires little elaboration. In this view, the self cannot be opposed, logically or existentially, to society, for it is only in relation to one's fellows that a person lives, through language, discussion, and feeling. Practical questions arise, however, about how groups are constituted and what governs relations between them. His outspoken presence in the movement to outlaw war illustrates Dewey's concern for this problem, but it remains a fluid concept in his work. He wrote in *Ethics* (1908) that "we cannot separate the idea of ourselves and our own good from our idea of others and of their good." How this commonality coexists with the concept of self-interested economic man is difficult to see, but even putting the problem aside, the notion of group identification begs the issue of group competition. For Dewey, in Robert Westbrook's analogy, society was "like a basketball team in which the different skills of the members of a team worked together for a common end." That end is, of course, the defeat of other teams; how do Dewey and Westbrook understand competition, fair and unfair? Who sets the rules, gets to play, or referees the contests?[33]

To his credit, Dewey appears to have lived in accord with the pragmatic code, in contrast to theorists who resist having their own concepts turned reflexively upon them. He accounted for many objections to his work, sharpened his ideas in light of counterarguments, and was not afraid to change his mind. This trait alone differentiates him from many intellectuals, and on the idea of social identity he performed a noteworthy modification. In 1939, at age eighty, Dewey rearranged his thinking significantly. "I should now wish to emphasize," he wrote in a piece called "I Believe," "more than I formerly did that individuals are the finally decisive factors of the nature and movement of associated life." He had witnessed the same rise of European and Japanese authoritarian regimes that motivated Walter Lippmann and others to rethink their positions on the place of social planning in a free society, and Dewey maintained his commitment to uninhibited inquiry by carving out a more elaborate place for the individual. Even so, his individualism never took on the rugged character that many imputed to Herbert Hoover, for the social character of society remained paramount.[34]

A commitment to democracy informed Dewey's entire intellectual project, but he was frequently confronted by antidemocratic interpretations of his words. Westbrook has recently devoted an extended study to this issue, and he makes substantial inroads into a systematic location of Dewey's aesthetics, ethics, pedagogy, and politics within a commitment to democracy. As of the mid-1930s, Westbrook shows, Dewey "was not arguing for 'social intelligence' as an alternative to politics," in a manner similar to people like Day or Hoover. Instead, Dewey began a forthright call "for a radical politics that incorporated social intelligence into its practice," a democratic adoption of the scientific spirit much like Beard's. Westbrook suggests that this political commitment, expressed through the League for Independent Political Action, *Common Sense*, and vigorous writing on many topics, superseded the reliance on rational consensus of which Dewey has been accused.[35]

In 1938, however, Dewey published *Logic: The Theory of Inquiry*. There he set forth a viewpoint that appealed to social scientists who sought to rationalize social relations. Once again preaching fusion rather than polarization, Dewey argued that multiple modes of reasoning were inappropriate to a nation facing serious challenges. The democratic society of free individuals needed to employ a common set of assumptions rather than continue to talk past one another. "The basic problem of present culture and associated living," Dewey wrote, "is that of effecting integration where division now exists. The problem cannot be solved apart from a unified logical method of attack and procedure." A monolithic logic, capable of being agreed upon by a majority of citizens, would insure a degree of consensus within competing constituencies, and the hope for a rational democracy is to be lauded as humanistic—at least in the abstract.[36]

Within the process of creating the collective logic that could solve social problems, Dewey either (a) erected an argument of unusual complexity or (b) engaged in outright contradiction. The same writer who called for a single mode of reasoning—logic—also wrote, in *Art as Experience* (1934), that "just as physical life cannot exist without the support of a physical environment, so moral life cannot go on without the support of a moral environment." Such a moral and artistic milieu would allow plural modes of reasoning. The reconciliation of logic and morality would be achieved, as in Charles Merriam's and other schemes of social engineering, by education: "The values that lead to production and intelligent enjoyment of art have to be incorporated into the system of social relationships." Dewey's sentence structure relies on a passive formulation that leaves responsibility—along with *intelligent* and *art*—undefined; *who* is to do such incorporating he never addressed.[37]

Dewey had long engaged in syntactic evasiveness, especially in regard to education. In 1898, he had argued for a sort of affirmative action; equality required that each individual "be provided with whatever is necessary for his realization, for his development, whatever is necessary to develop him to enable him to function adequately." Questions about the definitions of *realization*, *necessary*, and *adequacy* appear once again to be assumed away as commonsensical. Neither did Dewey consider the potentially paternalistic infrastructure implicit in the provision of these differential social advantages, perhaps meaning America to be an adult version of a Dewey school, where students' needs predominated. As Westbrook points out, however, the Dewey school centered more on the teacher than on the child, and some analogue of the directive teacher appears to have lurked within Dewey's conception of the logical democracy.[38]

Throughout his career, Dewey took aim at the problem of teleology and social goals. He was much more attuned to the problem of *how* to get things done—effectively and fairly—than that of deciding *what* to do. Fully aware of the challenges brought against him, Dewey could have been responding to Randolph Bourne when he wrote in *Reconstruction in Philosophy* (1920) that "when we take means for ends we indeed fall into moral materialism. But when we take ends without regard to means we degenerate into sentimentalism." A decade and a half later similar opinions persisted in a review of Lippmann's *Good Society*: "Definite and systematic exploration of the means, compatible with a free society of free human beings as the end, is, to my mind, the central problem, intellectually and practically, of genuine liberalism today." Because he worked from a variant of Ogburn's cultural lag theory, Dewey sought instruments of human expression that could maintain efficacy in the face in modern industrial dilemmas.[39]

By understanding humanity's context as contingent, Dewey created the need for a social method of coping with indeterminacy. He never offered certainty as a viable option; rather, the method of scientific inquiry could inspire sufficient confidence to face the future. By concentrating on means and methods, Dewey has worried many readers over the years, for the direction of those techniques was left unanswered. Once again he appears to have taken his critics seriously, for he explicated the matter most clearly in 1938: "The means have to be implemented by a social-economic system that establishes and uses the means for the production of free human beings associating with one another on terms of equality." But without a politics based on hermeneutics, how could those free human beings develop policies and institutions for the implementation of his ideal? Because scientific method filled a need more suitably than any other current option, he pragmatically invoked that method.[40]

People's questions about this problem often puzzled Dewey, who appears to have taken the ideal of human expressive freedom as a given for quite some time. He wrote in response to some criticisms by Lewis Mumford that "it would require a mind unusually devoid both of sense of logic and sense of humor—if there be any difference between them—to try to . . . set up a doctrine of tools which are not tools for anything except more tools." Dewey compressed modes of knowing, of contemplation, and of action within this variation of pragmatic thought: a conscious human could not know something without acting on the knowledge, and could not act without having known. Such puzzlement and irritation may have been symptomatic of the deeper problem in this argument. Of course everybody wants a free society of free individuals, he reasoned; how could any rational person want anything else? Like the rest of the rational reformers, Dewey never really acknowledged alternative systems of reason, grounded in ethnic, intellectual, and perhaps gender differences or in patterns of representation.[41]

These tensions between self and society, democracy and logic, and means and ends culminate in a reliance upon scientific method. The promise of science as only a provisional solution—but a solution nevertheless—to social problems appeared within Dewey's work for decades, but it evolved in response to technological and political changes. Optimism of a nearly irresistible sort inspired his reliance on science throughout his career; while he varied in his appreciation of the dangers of such an approach, never did there appear the crude technocratic dogma in which others could sometimes wallow. In *Individualism Old and New* (1930), he made his clearest declaration of scientific possibility. "Science is a potential tool," he wrote, "of a liberating spiritualization; the arts, including that of social control, are its fruition." The questions remain whether science as Dewey understood it resembled science as anybody else saw it, or whether science, widely and skillfully applied, could produce art.[42]

As a pragmatist, Dewey embraced a scientific ideal at odds with the purity of experimental investigation as an end in itself. Knowledge he again inextricably bonded to action, so that inquiry could only exist in relation to a problem to be solved. He held social science to the same imperative, dismissing people like Edmund Day who advocated the accumulation of pure scientific data that could then be applied by "social engineers." "It is a complete error," Dewey claimed, "to suppose that efforts at social control depend upon the prior existence of a social science." Instead, social science had to be energized by efforts to use it as a tool; his version of science mirrored engineering, not merely inquiry.[43]

This conception remains notably consistent throughout Dewey's writings of the 1920s and 1930s; nowhere does he allow for a science of society apart from efforts to alter the surroundings. The following quotations reveal this continuity:

> A new individualism can be achieved only through the controlled use of all the resources of the science and technology that have mastered the physical forces of nature.[44]

> What is sometimes termed "applied" science, may then be more truly science than is what is conventionally called pure science. For it is directly concerned with not just instrumentalities, but instrumentalities at work in effecting modifications of existence in behalf of conclusions that are relatively preferred. . . . Thus conceived, knowledge exists in engineering, medicine and the social arts more adequately than it does in mathematics, and physics.[45]

Dewey realized that commercial interests had co-opted these liberating methods of investigation, turning them into streamlined toasters instead of into truly social technology. He accordingly sought to enlighten a society of investigating citizens through education. Change could then be actively managed rather than passively experienced, the world shaped rather than consumed. The exuberance of his reasoning often goes beyond issues of methods and into the realm of ends. In *Individualism Old and New* Dewey asserted that "when we begin to ask what can be done with the machine for the *creation* and fulfillment of values" and "begin organized planning to effect these goals, a new individual correlative to the realities of the age in which we live will also begin to take form." Here we encounter the familiar sensibility once again, for the physical reality of the machine age inspired in many of its observers a confidence in applied science for the ethical, and sometimes metaphysical, fulfillment of the promise of the species (see fig. 9-4). In the end, Dewey insufficiently accounted for the presence of other forces in those artifacts; the toasters and airplanes represented not only engineering processes but cultural forces like exploitation and coercion.[46]

The logic of the machine process, as Veblen had called it, remains difficult to reconcile with a democratic politics unless politics no longer turns on the principles of free debate. Dewey's educational designs never fully accounted for the multiplicity of views of the world, a multiplicity that has been more acutely felt in the years since his death. Pluralism generates multiple phrasings of the relevant questions, not to mention competing answers to all questions. But the powerfully efficacious logic of applied science once again inspired social promise and also led an observer to underplay the role of epistemological politics in society. Tirelessly, Dewey attempted to incorporate science with democracy. In a revealing section of *Liberalism and Social Action* (1935), he argued that experts could enhance the overall social intelligence of a group. Attacking conventional individualism, he argued that within liberalism,

> native capacity is sufficient to enable the average individual to respond to and to use the knowledge and the skill that are embodied in the social conditions in which he lives, moves and has his being. There are a few individuals who have the native capacity that was required to invent the stationary steam-engine, locomotive, dynamo or telephone. But there are none so mean that they cannot intelligently utilize these embodiments of intelligence once they are a part of the organized means of associated living.

Such reasoning sounds appealing, but once again he begs the political question of deciding—by votes, markets, fiat, or other means—what values the political technology would embody. An entire population of political experts

Fig. 9-4. Interior, Norris Dam: power as sterility (Tennessee Valley Authority)

may not be required to rationalize a democracy, but for Dewey to imagine experts who would design the overall set of priorities ignores the sense of self-interest on the part of those in power.[47]

Dewey was not a technocrat in any usual sense of the word, for he explicitly denounced aristocracies of the competent throughout the 1920s and 1930s in *The Public and Its Problems*, "I Believe," and elsewhere. But how likely was it that citizens would uniformly embrace the engineering mind or that a vanguard of political and social engineers would design a rational democracy? As Richard Rorty has written, "I think Dewey was at his best when he emphasized the similarities between philosophy and poetry, rather than when he emphasized those between philosophy and engineering." Because, like Veblen, Dewey tied social improvement to engineering modes of reason, he forced a confrontation between competence and hierarchy on one side with inefficiency and pluralism on the other. Citizens continue to long for democracy that works, but by making reason an unarguable category instead of a contested judgment, he perpetuated the prospect of an apolitical scientific politics. Even while his own belief in democracy was steadfast, Dewey provided logical justification to less democratic writers with whom he disagreed, and this responsibility for their aid and comfort remains one aspect of his legacy.[48]

Some of the most powerful of those intellectuals found work in the New Deal administrations of Franklin Roosevelt. In contrast to the figures in this chapter, many of whom wrote with more complexity and less influence, academics such as Wesley Mitchell, Charles Merriam, and Rexford Tugwell used the emergency powers of the moment to attempt to institutionalize rational reform's basic precepts. Their efforts and eventual rebuke, like the undefined aspects of contestation in Dewey's work, once again pose the paradoxes inherent in any attempt to conquer politics with apolitical reason.

10

SOCIAL ENGINEERING IN THE DEPRESSION

II: INSIDE THE NEW DEAL

The hectic atmosphere of the New Deal allowed many new issues onto the national agenda. Herbert Hoover's semivoluntaristic conceptions of planning suddenly looked so timid that much more thorough-going programs of centralized responsibility became tenable. In the first years of Franklin Roosevelt's presidency, the advisors who wanted comprehensive public-sector planning instead saw FDR try to institutionalize quasi-private organizations, and in so doing endorse a modified Hooverian associationalism. That experiment in foxes guarding the economic henhouse —the National Recovery Administration—failed quickly and dramatically, discrediting other conceptions of social engineering in its demise.

By 1933 the idea that America needed plans, and not simply unimpeded market relationships, had become widely accepted. Planning proponents paraded documents from the Mayflower Compact forward to validate the notion with historical precedents. Cities, national businesses, regional groups, trade associations, academic professions, philanthropies, and other entities recognized the need for coordination, foresight, and deliberation. Two concomitant issues, however, inspired controversy. First, whose plans would be implemented? Gerard Swope and his brethren called upon an enlightened business leadership to point the way with minimal interference from labor, government, or consumers, while many other proposals put theoretically apolitical experts at the controls. Second, what was the goal of a planned community or nation? This problem of orientation handcuffed Hoover and Roosevelt, for the competing visions of collective purpose precluded any concentration of techniques on a common objective. The experience of war inspired the most dramatic rhetoric from Roosevelt, who worked to establish his right to plan and to organize, rather than in more purely Keynesian fashion to spend. Government became a player in the quest for power and not merely a trophy to be won.[1]

This conflict between the business version of planning, which many argued had been given a more than adequate trial before 1929, and the rising call for activist government animated much of the debate over planning during the 1930s. While business executives saw their plants, mines, and stores as the key to the nation's economic health, the administrative outlook of the new liberal planners led them to think of this economic infrastructure "not as private property but as public utilities," in the words of one historian. While this difference in outlook would see its most dramatic unfolding in the struggle of the Tennessee Valley Authority to buy electric companies, the New Deal administrators busily tried to orchestrate many aspects of national life within and beyond the economic realm; crop rotation, land use, internal migration, race relations, recreation, and electric power came under bureaucratic control. Within the indistinct confines of such a fragmented movement, isolating the role of engineering models in the planning programs is impossible, especially because of the ways that engineering had informed social science and managerialism in the 1920s. Roosevelt's "positive state" intervened in American life in so many ways that teasing the technocratic strand out of the fiber contributes only partially to understanding the New Deal.[2]

In the midst of the considerable inaccuracy that surrounded the term *planning* in the 1930s, various connotations gained credence while others faded. The economists closest to the social engineering sector took care to avoid hyperbole while retaining the positive connotations of science. Such economists—Mordecai Ezekiel, Lewis Lorwin, and, to a lesser degree, Gardiner Means—stressed the need for conscious coordination of the American economy. They pointed to the failures of presumably unregulated markets with their "readjustments which previously were supposed to be brought about automatically without anyone thinking about them," according to Ezekiel. Science, in contrast, would enable rational discussants to reconcile their differences and move purposefully forward; social arrangements, in the continuation of the old argument, would catch up with technical development. In Lorwin's words, "Parliaments try to resolve these conflicts [between labor and capital, agriculture and manufacturing, and creditors and borrowers] by compromise. Dictatorships meet them by usurping the right and power to impose a solution by governmental decree backed by force. A system of economic planning aims at resolving such conflicts by working out long range national objectives and by the use of research and scientific analysis." Escaping politics with an elegant abstraction rather than within history had become a familiar theme in groups like Mary van Kleeck's IRI, at whose conference Lorwin spoke.[3]

When his chance came to implement a program of centralized economic coordination, Roosevelt set up the NRA. It appeared to satisfy simultaneously the social engineers (who would monitor adherence to administrative codes rather than price and profit figures) and the business communities, whose domination of the code formulation process often gave them a federally sanctioned monopoly pricing structure. The NRA failed for many reasons, among them public dissatisfaction, price gouging, and unconstitutionality. For rational reformers, the problem of accumulating adequate, current, and complete data from which to plan came home yet again as it had in World War I. The American economy was simply too complex, interconnected, and diffuse for bureaucrats to mandate prices in the absence of market forces. Lorwin later estimated that a full decade of gradual phasing in of codes, one industry at a time, still would have been insufficient; the NRA, as it was, attempted in vain to codify over five hundred industries overnight. Producers restricted output to maintain prices, putting more workers on the street and further slowing recovery. The entanglements of code adherence penalized businesses large and small, so when the NRA died in 1935, no major constituency criticized the Supreme Court's verdict in the Schechter case. Everybody appears to have had enough.[4]

Because of its inclusive character, the NRA gave opponents of both business associationalism and self-policing, and governmental planning and administration substantial ammunition. One NRA official recalled the experience as a horrific nightmare; it "was characterized by a lack of definite policy and of proper understanding of objectives. . . . Various economic panaceas equally or more contradictory were borrowed from long agitated movements, both good and bad, and hastily thrown together into an ensemble of contradictions." When the economy failed to respond to such inept ministrations, critics declared the whole conception of rational administration to be tried and found wanting: the remaining planning programs in Roosevelt's arsenal all suffered from the credibility sinkhole that was the NRA. The aberrant behavior of director Hugh Johnson did nothing to increase confidence in nonmarket price mechanisms, bureaucratic coordination, or economic forecasting.[5]

Some of the planners responded with bitterness, attacking the old bogey, politics, continuing to believe it could be overcome by logic. George Galloway, a sometime government planner, insisted that he and his kind could take a "detached, anthropological attitude toward public affairs" in order that national economic planning could do for America what scientific management did for single firms. Politicians, meanwhile, recognized such rationality as a threat to their role as "brokers and 'moral midwives' between pressure groups seeking special privileges." In such an understanding, capitalism and democracy clashed. His call for a soviet of technicians may have lacked the drama of Veblen's, but Galloway and his kinsmen put no less faith in the social application of the engineering method.[6]

While the NRA did little but give hard-core planners a brief and unfulfilling taste of what real governmental commitment to the social engineering model could be, other efforts within the New Deal labyrinth did employ the theories and personnel of the pre-Depression rational reform enterprises. Because of his pragmatic and ideologically fragmented approach to government, Roosevelt made no concerted attempt at scientific reform. Outposts of planners, administrators, and other veterans of the social engineering movement in the 1920s did exist, however, in three offices in particular: the NRPB, the TVA, and the Department of Agriculture. All were ultimately defeated by the same internal dynamics of log rolling, compromise, and power plays that science was alleged to make obsolete.

THE NATIONAL RESOURCES PLANNING BOARD

In addition to the NRA, another less publicized aspect of the National Industrial Recovery Act of 1933 was the creation of the National Planning Board. Its mutations revealed evolving conceptions and expectations of federal planning as changes in the board's responsibilities and lines of authority responded to criticisms from legislators and other observers. The NRPB attempted to bring rationality to the expanding presence of federal intervention, beginning with public works spending, while also recognizing the realities of public and congressional opinion. By no means can the National Planning Board and its successors be viewed as bastions of hardheaded technocrats, but engineering continued to fuel aspirations toward scientific government in the writing produced by the board and its members. The board left a mixed legacy. As Barry Karl argues, it resisted stereotypes common to much of the New Deal: it was not merely benevolent pragmatism; it did not centralize authority in Washington; it was not "freewheeling"; and it did not employ a rapidly hardening bureaucracy of full-time administrators. The NRPB also did not have the impact on American life its supporters imagined it might.[7]

Among the key personnel involved in the NRPB, several—Frederic Delano and executive officer Charles W. Eliot II—are best considered as members of a publicly minded gentry who had little time for exaggerated conceptions of scientific capability. Delano, the chairman of the board during its first and last years of existence, called himself a conservative liberal and put forward a notion of planning so broad as to offend no one. Delano appears to have been accommodating and collegial, although he had limited stamina for bureaucratic infighting. He also consistently deflated the NRPB's ambitions. As he told Stacy May of the Rockefeller Foundation, "no single organization can, or should, attempt to do this [planning] generally throughout the country." Sitting on the board of the Brookings Institution and being involved with urban planning for many years gave Delano a respect for the limits of his post.[8]

Some of his colleagues on the NRPB had broader ambitions. Charles Merriam and Wesley Mitchell, and two "advisors" to the board, Beardsley Ruml and Henry S. Dennison, brought to Washington well-formed conceptions of experts, governance, and administered economies. While the NRPB was kept out of the politics of policy formulation, in large measure by the board members' wariness of experts flouting the democratic process, the NRPB itself showed flashes of its managerial and technocratic heritage—but only flashes. In their own separate writings of the period, however, Merriam, Mitchell, Ruml, and Dennison continued to struggle with the same concepts they had in the 1920s. The experience of Washington politics appears to have chastened some of the more confident advocates of centralized social scientific control even as the scope of national problems made more apparent the insufficiency of localized action.

By one count the NRPB produced 370 economic, land use, and social studies. This staff distinguished itself as it included the future Nobel Prize winners Milton Friedman, Wassily Leontif, and Paul Samuelson, along with John Kenneth Galbraith; the studies were carefully and often expertly done. In terms of policy and institutional philosophy, one document among the reams of material may be the most important: the 1933–34 final report of the National Planning Board, also titled "A Plan for Planning." Here, early in the board's life, Delano, Merriam, and Mitchell attempted to spell out their

Fig. 10-1. Frederic Delano, 1932 (Wide World/AP; Franklin D. Roosevelt Library)

sense of how science, democracy, and resource utilization could fit together. Throughout the document, admiring references to managerial and engineering technique exhibit a confidence in rationalized control that would soon be mangled in the gears of an intractably complex American economy and society.[9]

Mitchell and Merriam brought to the NRPB a powerful awareness of the precedent set by *Recent Social Trends*. The problem, however, soon became clear: if the board's function was to be more than purely investigative and less than unaccountable, where did it fit? Roosevelt and Harold Ickes, the secretary of interior, had to keep moving the board from niche to niche to keep it alive. Powerful patrons appeared to take planning seriously even as the concept came under nearly annual attack by legislators. The choice facing Mitchell and Merriam echoed from their days at the SSRC and the NBER: where did expert detachment (and its distance from the often hypnotic centers of power) end and political advising (and its associated rewards and hazards) begin? "A Plan for Planning" attempted to maintain the possibility of a rationally ordered economy and to avoid the appearance of bureaucratic elitism by appealing to tradition and to scientific models of managerialism.

Like many pro-planning books and articles of the time, the NRPB document interpreted any attempt at foresight in American history as a comforting precedent for "planning," an idea "as old as the Constitution and as widespread as business enterprise." In fact, argued the unsigned booklet, "the Constitutional Convention was a large-scale planning board." The writers wove this history into the American tradition of inventiveness so that modern-day planners became the lineal descendants of the Yankee tinkerer and the town meeting legislator. Such spiritual strength as only applied industrial science could provide would rescue America from its present malaise. "When we are resigned to drifting and too weary to plan our own American destiny, then stronger hands and stouter hearts will take up the flag of progress and lead the way out of difficulty into attainment," read one section of the document that sounds as though Merriam wrote it. The spirit of material accomplishment and record of success compiled by engineering clearly suggested a ready referent for those searching for stability and success.[10]

The "Plan for Planning" stands as a monument to positivism that would soon topple. "With their research in the field of human behavior," social scientists, steadily drawn in parallel to natural scientists, "should correspondingly facilitate the making and perfecting of social inventions." Harlow Person, Stuart Chase, George Soule, and the other advocates of Taylorite and Veblenian productivity also saw their heroes invoked. "What stands between us and the realization of the hopes that gleamed before the eyes of our people from the earliest days," the report claimed, "are only our attitudes and our social and political management." Such confidence in method indicates not a single-minded reliance upon a clumsy analogy between government and engineering but rather an attempt to tap any sources of methodological or moral energy that would sustain the planners in the face of their critics.[11]

The NRPB writers also added one other element to their argument. The objectivity of scientific method, for so long an element of faith among technocratic progressives and liberals, combined with the board members' independence from electoral politics to make them, in their own argument, powerful advocates of some undefined public well-being. In this reading, opposition to the planners often indicated self-interested designs. "It may be found," the report claimed, "that some of those who cry 'regimentation' when public planning is mentioned foresee interference with their own practices of private regimentation and exploitation." The promise that managerial attitudes and scientific standards of objectivity could overcome politics became a nightstick with which to bash sometimes legitimate self-interest. The imagined authority of the applied scientific image, however, could become an imperialism in and of itself.[12]

Because work for the NRPB was only part time at the highest levels, Dennison, Ruml, Mitchell, and Merriam continued to write on planning topics under other auspices. While equating their private or scholarly writing with NRPB policy is clearly erroneous, their work helps to show some of the attitudes present on the board, whether or not they come through in its documents.

Dennison wrote little in the mid-1930s, perhaps indicating a reevaluation of the stands he took in earlier managerial writings. Ruml joined the NRPB, with Dennison, in an advisory capacity in 1935 and then replaced Mitchell, who resigned in September of that year, on the board itself. His advisory memos to the Rockefeller Foundation and to the NRPB revealed a continuing yet evolving commitment to rational reform. His rhetoric still stressed "the advancement of social understanding and control," but no longer did the social scientist have pride of place. By 1938, Ruml wrote little on social science and instead gave considerable thought to the role of government spending in the maintenance of purchasing power. As a businessman at Macy's he continued to maintain independence from Swope's or Dennison's advocacy of private initiative, and he concentrated on the largest levels of national policy. During the Second World War he devised a new income tax collection scheme that briefly put him into the national spotlight.[13]

Mitchell served on the NRPB in the midst of both familiar assertions and growing doubts about scientific reform. Throughout the early and mid-1930s, he continued to tie together engineering, managerialism, and economics as he envisioned a rational method of social change. At a New York University conference in 1932, he said that "not only the economists, but also the engineers, the business men, and the public servants [the university] trains must cooperate in the scientific discoveries, the practical inventions, the controlled experiments, the routine administration which is called for." Such a method, in contrast to "the dangers of 'inspirational reforms'" and "reforms that produce almost as much harm as they remove," would improve human welfare through the elimination of the losses sustained in chasing red herrings, pointlessly debating some point or another, or operating out of ignorance.[14]

Engineering's mastery of the physical universe continued to serve for

Mitchell as a dramatic example of what social science could attain. Because natural scientific logic underlay the technological innovation that taxed traditional structures of knowledge and action, people needed a scientific social attitude. He told an NBC radio audience that "if men are ever to attain a degree of skill in dealing with social changes comparable to the skill they have attained in turning natural resources to their use"—taking such a goal as a given—"what we need to guide public policies is a similar development of the social sciences and a similar application of social science to practical affairs." Social process could be logically cumulative, just like applied science. (See fig. 10-2.) In the same way that "engineering technique has reached its high state of efficiency" by each generation's being able to stand on the shoulders of the last, "so we must expect that the development of an efficient economic technique will require a long series of discoveries, inventions, and practical trials." After witnessing the NRA debacle and experiencing firsthand the struggle between expert data gathering and political advising, he eventually softened his rigid analogy between social engineering and the original kind.[15]

Mitchell often couched planning in terms of "the attempt to use intelligence as the guide to action," like Dewey distancing himself from the authoritarian scientism of technocracy. By 1938 the annual report of the NBER no longer confidently predicted scientific direction of society. Instead, Mitchell admitted that "there is no assurance that economics can be made to give satisfactory answers to all the practical questions that face us as citizens." But, he continued, "what other effort to enhance human welfare has a brighter promise in the long run than the application of scientific methods to social problems?" One of his economist colleagues agreed that Mitchell had to reconsider the base of so much previous confidence. "In 1923," wrote Alvin Hansen, "he felt pretty sure of himself and was ready to make positive recommendations. Twenty years later, that was no longer the case." Instead, economics appeared to be so complex that increased knowledge prompted uncertainty, not confidence.[16]

More than any other NRPB principal, Charles Merriam struggled to balance democracy, education, efficacy, and social science in his writing and administration during the Depression. Because of his experience in Chicago ward politics during the 1910s, he knew the stakes and tactics of administrative power plays once he got to Washington. Even so, his commitment to scientific objectivity, a value he held high, overruled any desire he might have felt to mix it up with the NRPB opponents. Instead, he stressed logic and reason as he sought to convince the skeptics and opponents that plan-

Fig. 10-2. Frontispiece to F. Stuart Chapin, Cultural Change *(New York: Century, 1928). It bore the caption: "The Chicago Tribune Tower. This Beautiful Structure, Representing a Modern Skyscraper Made of Reinforced Steel Construction, with the Gothic Motif Superimposed, Epitomizes and Summarizes the Chief Theory of This Book—That Cultural Change is Primarily Accumulative."*

ning could be American, effective, and democratic. Like Mitchell, though, Merriam left the NRPB with less confidence in the program of rational reform than he came with in 1933.

As he had throughout his career, Merriam needed science to inform education rather than to certify elites, so few engineering references appear in his New Deal writing. "The task of politics in our day," he wrote, "may be stated as 'the translation of scientific gains into social gains under the direction of human intelligence.'" While he underlined the NRPB's commitment to "make the fullest possible use of scientific methods in dealing with our national resources whether natural or human," Merriam also insisted upon democratic participation: scientific intelligence had to be social for government to be democratic. He was more an apostle of planning as a process than an advocate of particular schemes, of which he had few. Even in the "what if" stages of discussion, he made the public whose lives would be part of the planned environment part of the process: "Assuming that adequate social engineering can be found *and can be supported by the masses with whom the ultimate decision lies*, a new world is well within our reach if we can organize and act to attain it." He maintained his faith in mainstream planning as distinct from the extremist designs and demagogic radio broadcasts put forth by Huey Long, Father Charles Coughlin, and Francis Townsend. In light of such challenges and others made possible in part by new technologies, "sneering at intelligence in human affairs," he said in the same address, "is defeatism of the darkest type." Like Dewey, Merriam tried to use science to repair politics, not transcend it.[17]

Because of this commitment, Merriam sought to decentralize the planning function, giving regions and cities viable roles to play in data gathering and idea formation. Furthermore, he avoided political infighting that could have kept the board alive, preferring not to lobby legislators or confront turf-conscious bureaucrats. Afterward, he could "not recall any instance of partisan politics entering into any of the many conferences and discussions" of the board. The same staff of experts and decentralized advisors of which Merriam was justifiably proud, however, went virtually unused as a resource for the board's survival. The NRPB could have no official role in policy formation if its scientific integrity were to be preserved, but Merriam did little to circulate its findings. In meetings, NRPB officials remained silent unless asked a direct question by a legitimate policymaker so as to avoid any hint of advocacy. Such purity of self-conception doomed the NRPB in a political atmosphere as highly confrontational as Washington. Instead, Merriam continued to proclaim his board's function in the lofty terms of technologically empowered humanism. He wrote that "in government, as in industry, the strategy of the new scientific and technological world calls for . . . less of authoritarianism, and not more; less charity and more fraternity; less shooting and more persuasion; less drifting and more consideration and planning of objectives." But as he discovered, proclaiming unchallenged hopefulness soon became impossible.[18]

ARTHUR MORGAN AND THE TVA

In contrast to the NRPB and many other agencies, the TVA attempted both civil and social engineering. Its significant technical achievements became cultural symbols of southern renewal, government efficacy, and architectural monumentalism. More important here are the extensive plans for political regeneration developed by the head of the TVA, Arthur E. Morgan (1878–1975). An admirer of Edward Bellamy and of many better-known rational reformers—Edwin Gay, Walter Lippmann, Henry Dennison, and John Dewey—Morgan brought a panoramic vision of civic reconstruction to his post. In the end, he was exiled from the realm he attempted to reform by a peculiar combination of moralism, mysticism, and scientism. He was also the victim of astute administrative politicking by his associate David Lilienthal.[19]

Morgan served as the president of the experimental Antioch College in the 1920s, where he educated people to become "philosopher-engineers." He supported many aspects of the rational reform spectrum, joining the American Eugenics Society as a charter member and holding that Edward Bellamy's socialized rationality showed significant promise for the alleviation of industrial problems. (Yet at age eighty-seven, Morgan marched with others at the college to show solidarity with Martin Luther King, Jr., who at the time was at Selma.) Once he got his chance from Roosevelt, Morgan worked quickly yet naively to institutionalize his design for a better world. Soon after assuming his post, he drafted an extensive ethical code for TVA employees, a document that put him at odds with Lilienthal and the other director, Harcourt Morgan (who was no relation). Lilienthal resented the imposition from the top down of a plan by "supermen," and he steadily attacked Arthur Morgan from within the TVA structure, fearing Morgan's incompetence and enhancing his own political future. Appearing before Franklin Roosevelt in an extremely unusual administrative hearing in 1938, Morgan was accused of making false and unsubstantiated charges against Harcourt Morgan and Lilienthal in connection with their efforts to buy

Fig. 10-3. Hiwassee Dam. According to one historian of the TVA, "The great ability of the Moderne Style to suggest an exciting future (the twenty-first century?) is especially evident on the projected deck with all its contrast and glow" (Walter Creese, TVA's Public Planning *[Knoxville: University of Tennessee Press, 1990], p. 222). (Tennessee Valley Authority)*

power-generating facilities from Wendell Willkie's Commonwealth and Southern Company. In the bizarre hearing, Arthur Morgan refused to answer Roosevelt's questions, preferring to read prepared texts, which often had little connection to the charges. FDR had no choice but to dismiss Morgan in March 1938.[20]

The ethical code and Morgan's philosophical writings, in particular a book entitled *The Long Road* (1936), play out the reasoning behind his unconventional interpretation of the planning impulse. In all instances, he insisted upon the primacy of moralistic rectitude as the raw material of social engineering. He wrote that "one might compare our capacity to create a better human society to our ability to build great suspension bridges." In such a comparison, "we may liken personal character to the quality of steel of which a bridge is made." Probity of this sort, once inculcated, would transform social intercourse: it would "solve seemingly inextricable complexities; [and] make unnecessary and meaningless vast systems of checks and balances, of laws, regulations, surveillances, inspections, and prohibitions." Personal moral strength could discipline drives, restrain antisocial actions, and, in Bellamyesque fashion, render politics obsolete.[21]

In such an antipolitical society, all those citizens of good character would not migrate toward full equality. On the contrary, "consensus of judgment would not mean taking formal votes on the 'one man, one vote' principle. Consensus of judgment may be arrived at by the deference of the many who do not know to the superior judgment of the few who do." This philosophy seemingly relies on the same positivism that underlay the one best way within Taylorism, but the latter's technical component diminished in Morgan's outlook. His reliance on the self-evident nature of social objectives, and the resulting faith that reason could transcend politics, remains consistent with Rockefeller philanthropists and other reformers. "The Tennessee Valley Authority," read the TVA ethical code, "should be unusual in the same way that a great engineer is unusual, by carrying into effect with unusual thoroughness and courage *the principles of conduct that most well intentioned persons are agreed upon as being desirable.*" Morgan's belief in the quiet acquiescence of the many who do not know derived from his notion of character; he wanted to "lead each person of his own volition to try to play that part which is best for society as a whole." Unlike other social engineers, Morgan left *character* defined in traditional terms and did not endorse scientific citizenship. "An unusual engineer does not discover new fundamental principles of physics," he said by way of analogy, "but is unusually successful in putting into practice principles already known." For Morgan the social equivalent of such agreed-upon commonplaces was a blunt utilitarianism reinforced by some generic reliance on the Golden Rule: "Ethical conduct is simply that conduct which is best in the light of its total consequences."[22]

To regularize TVA employees' understanding of such principles, the ethical code sought to teach unsurprising moral lessons not far removed from "A scout is honest." A steadfast refusal to behave politically motivated Morgan's entire document, for he explicitly forbade many actions and practices that had become normal in the distribution of as much pork and power as the TVA represented. That enterprise was not only to serve as a rate "yardstick" with which to compare private power companies, but a moral one as well: "If, as a result of our effort, the Tennessee Valley should become the richest and smartest part of America, but if in getting that result we should leave an example of deceit, exploitation, favoritism, patronage, extravagance, bad personal habits, and selfish personal ambitions, our efforts might do more harm than good."[23]

In concrete terms, Morgan left no sin unanticipated and expanded a long tradition of American corporate paternalism into government. He called upon the TVA employee to refuse lunches bought by any potentially interested party, refrain from interdepartmental rivalry, and "welcome someone going beyond him [by means of promotion] if that person deserves it, or if the good of the service requires." Business affairs were to be conducted in the open; no tips were to be accepted; and, in the purchase of land from reluctant sellers, "dickering and bargaining should not be introduced, but arbitration or condemnation would be in order." Personal habits also fell under TVA purview, for "dissipation and other habits which destroy health and the full possession of one's powers are in direct conflict with any reasonable ethical code." Similarly, alcoholism, "lax sex morality, gambling, and the use of habit forming drugs" were frowned upon, while "friendliness . . . is essential." Significantly, the code contained neither sticks nor carrots: enforcement was ignored, as if the simple and unarguable reasonableness of the document would insure compliance.[24]

Morgan's utopian hopes doomed him in the volatile political setting of the New Deal, but the TVA became a unique and important node of planning activity in the 1930s. It brought together regional planners, large-scale national planners like Rexford Tugwell and Stuart Chase, and technocrats who focused on the primacy of energy in the economy. The soaring shapes of the dams' futuristic gantry towers juxtaposed symbolically with efforts to use technological reasoning to reconfigure an entire way of life; designers of generating systems worked with planners of communities under often comple-

*Fig. 10-4. Kentucky Dam: civil engineering as a metaphor for civic
engineering (Tennessee Valley Authority)*

mentary suppositions. Morgan, furthermore, attempted not only to reshape
the South of the TVA; he envisioned his utility agency as a working model for
a national program of technological development, political renewal (or es-
capism), and resulting social harmony. The phrase attributed more commonly to Daniel Burnham—"Make no little plans; they have no magic to stir
men's blood"—fits Arthur Morgan just as aptly.[25]

REXFORD TUGWELL

In the person of Rexford Guy Tugwell (1891–1979), several planes within
social engineering converged. No advocate of actively technocratic reform
rose to a more powerful position, and no one was better connected: Tugwell
taught with Wesley Mitchell at Columbia, knew Mary van Kleeck and
George Soule from the Taylor Society and elsewhere, and worked closely for
Edmund E. Day in the preparation of a study of education's role in the transformation of American society. At the time of William Ogburn's death in

1959, the two veterans of the social science boom years were working on a book of social theory for a technological society. After trying to reorganize America's farms, Tugwell served as a city planner in New York, with poor results, before trying, once again with limited success, to govern Puerto Rico with the same ideology. He drew heavily on Taylor, Veblen, and Dewey in the development of a completely mechanistic social theory that centered on production of consumer goods. Tugwell also unreflectingly peppered his many speeches and writings with admiring metaphoric and analogical references to the rigid and apparently efficacious logic of engineering. Social engineering had begun to come under heavy criticism by 1935, and the reincarnation of social engineering of a different yet related sort after World War II involved new institutional actors, political languages, and social theories. When Tugwell left the New Deal after the prevailing winds shifted, no other figure stepped into the void; he was the last and most flamboyant of the prewar social engineers.

As Ellis Hawley has demonstrated, the Department of Agriculture "had developed in the 1920s in such a way as to be ready with economic planning vocabulary and apparatus when state intervention became possible" in the New Deal. Because Hoover had avoided a parallel course in industrial affairs, leaving such matters to the NBER especially, the apparent paradox of an industrial technocrat going to work with farmers makes more sense; Tugwell went to the department most ready for his approach. His theory combined the order and rigidity of the Taylor system of industrial management with Dewey's insistence on social experimentation; in his memoirs Tugwell recalled thinking that "a Taylor was needed for the economy as a whole," while elsewhere he wrote that the New Deal was best described as "a charter for experimentation, for invention and learning," Deweyan concepts all. Ignoring both his mentor Simon Patten—who insisted that the firm and the society operated differently—and Taylor, Tugwell wanted to administer the whole of America in terms of rationalized industrial productivity.[26]

The goal of control became an end in itself for Tugwell. While an engineered environment could let people live psychologically full lives, in a complex technological world the masses failed to recognize that fullness when they saw it. Instead, the experimental dimension in such a period had to be provided by experts blending Taylorite centralization with Deweyan tests of social utility. Veblen's matter-of-fact thinking, the cornerstone of what Tugwell called the industrial discipline, thus found full expression in Tugwell's insistence upon ahistorical technical expertise as a replacement for an apparently inadequate politics of democracy and virtue.[27]

Fig. 10-5. Rexford G. Tugwell, 1933 (UPI; Franklin D. Roosevelt Library)

Tugwell had a fully formed outlook in place by the early 1930s, when the General Education Board of the Rockefeller Foundation engaged him to study "social objectives in education." Such an investigation resonated with the many social engineering inquiries undertaken under Rockefeller direction. As Charles Merriam saw, public education in what proponents called scientific citizenship lagged considerably behind the most advanced methods employed by leading social scientists. Seeing just what social values *were* being propagated made logical sense, and the study had indefinite and ambitious outlines, again much like other Rockefeller projects. In the report to the board, Tugwell relied on a strict analogy not only between social science and natural science, but also between the industrial firm and the industrial nation: "What is the content of this term 'managed society'? In general the future is a technological problem." Given this finding, the solution existed within clear administrative and political value systems: "Considerable technical background would be needed to approach [social planning] intelligently. . . . [But] democratic interference in a determination would involve a *dangerous substitution of vague desires for expertness*." The expert most central to this new regime would be the economic researcher, who would "marshal the resources of intelligence and techniques of management and administration" to replace laissez-faire with purposeful control.[28]

In *Redirecting Education*, the published version of the report on education, Tugwell expressed puzzlement that the idea of social management could be at all troublesome: "Why ought we to be logical, scientific, and rational in other areas of life but not in social affairs?" Tugwell expressed his views in a study that appeared as usual without any attribution to the Rockefeller Foundation agency whence it originated. Day's engagement of a leading social engineering advocate—he told Tugwell that "you do not need to be told that I am much pleased with the developments reported in your letter of January 20 [1932]" in which Tugwell accepted the assignment—reveals as much about that institution's philosophy as about the economist.[29]

Among Tugwell's other writings, perhaps his book of 1935, *The Battle for Democracy*, most completely reveals his technocratic tendencies. The very title is an exercise in irony, much like destroying a Vietnamese village to save it: he no more wanted to increase democratic participation than he wanted to hand America over to the cartelists. His reasoning had not changed from before. Because capitalism of the Adam Smith variety was a myth, explicit measures had to be consciously created. "The jig is up. The cat is out of the bag," he wrote. "There is no invisible hand. There never was." Instead, "we must now supply a real and visible guiding hand to do the task which that mythical, non-existent, invisible agency was supposed to perform, but never did." The visible hand would be that of applied science, the same force initially responsible for social dislocation. "The wounds made by applications of science can be healed only by a further extension of applications of knowledge and intelligence," he argued, ignoring the possibility that these actions could simply exacerbate the injury.[30]

In *Battle*, Tugwell located the fullest development of scientific reform among social scientists. He called on scientists, economists, and managerialists to reshape the world; democracy had very little to do with his theory. He proclaimed it "a kind of duty among civilized beings now not to desert reason but to press its claims insistently," with the handy implication that those who opposed reason of the sort he favored were uncivilized. To the new master of the natural and political world, Tugwell's call to rational action was louder yet: "The scientist who abhors meddling with society," he claimed, "is altogether too naive." Contrasting such ostrichlike opponents, he and his fellows saw themselves as modern reformers, hardheaded and effective, bearing the American tradition of efficiency in industry, government, and society. He made this connection most tightly in *The Industrial Discipline and the Governmental Arts* (1933), a book praised by none other than John Dewey, who called it "by far the most intelligent analysis of our present economic situation and its impact upon the social order that exists."[31]

Such reliance on presumably self-evident technical reason led Tugwell to make some noteworthy innovations while he was in government. He and his former student Roy Stryker developed the Farm Security Administration photography project along self-consciously scientific lines; its great historical value runs in direct opposition to Tugwell's ahistorical scientism. On the other side of the ledger, to this day the ham-fisted tactics of the Resettlement Administration are, in some rural areas, still recalled with the utmost bitterness by citizens who resisted "experts" telling them they had to move. Perhaps one article from 1940 tells observers much of what they need to know about Tugwell. He argued for an area of government, free from the responsibilities of checks and balances, where purely logical authority could originate. By his own experience, though, Rexford Tugwell could never escape the responsibilities and challenges of disagreement and compromise: for him, as for the other social engineers, there was no "superpolitical" place to escape the illogic of politics.[32]

. . . .

The transition in the New Deal just after Roosevelt's first reelection relegated the planning impulse and the aspirations of many social engineers to the slag heap of failed ideas. Even though the Blue Eagle codes were hardly a fair test of nonmarket control of the price system, the NRA served as an irritating symbol of how far the advocates of centralized economic control had to go to develop their techniques. The regimentation of the codes and the scope of Roosevelt's experimentation, played out against a backdrop of advancing

European political crisis, did increase criticism of the idea of social engineering itself, for the supposed virtue of the scientist was proving itself inadequate as a civic ideology; Stalin, Hitler, and Mussolini turned scientism to much more dreadful ends in a matter of years. By 1939, many American intellectuals openly challenged the rational reform program. They examined technocratic language more closely and found it to introduce into politics some stowaway notions incompatible with belief structures, that of freedom in particular, that still summoned strong emotions. Similarly, the networks of social engineers—including the SSRC, Rockefeller philanthropies, managerialists, and planning apostles—diminished in importance. As doubt grew and the oncoming war demanded more and more attention, it became apparent that applied scientific logic alone could not rescue an America seemingly dead in the water.

PART FOUR ·································

RECONSIDERATION AND RETREAT

1 9 3 4 - 1 9 3 9

Of all the sorrows that afflict mankind, the bitterest

is this; that one should have consciousness of much,

but control over nothing.

HERODOTUS

11

RECONSIDERATIONS

Resistance to the colorless regimentation connoted by the phrases *machine age* and *machine civilization*, originating in the 1910s, grew during the next decade. But not until the 1930s did a sustained rethinking of social engineering appear. When the challenge did come, opposition quickly mounted to technocratic, social scientific, and managerial control of American economic and social life. In a transatlantic political context, many intellectuals—some formerly committed advocates of scientism and its associated possibilities—began to compare technocratic reform with the marching dictatorships that often validated their domination with science. In addition, the failures of the NRA, the excesses of technocracy, and the quickly bloated and unaccountable bureaucracies associated with administrative control made many observers wary. In these reconsiderations, however, an uneasy relation to technological innovation and development persisted, for no writer could solve the problem of reconciling the body politic with the ghost in the machine.

LEWIS MUMFORD AND *TECHNICS AND CIVILIZATION*

Lewis Mumford (1895–1990) made perhaps his most lasting impression on the world of letters in 1934 with the publication of *Technics and Civilization*. A sprawling, suggestive, and synthetic work, the book attempts to comprehend, and in so doing turn to human purposes, the machine process. While the book ultimately exhibits crippling internal contradictions and fails to provide for humanistic control over the made and making environments, Mumford did raise major questions that many rational reformers neglected.

A noble failure (at least in part) rather than a safely conventional success, *Technics and Civilization* remains valuable for its aspirations and insights.

In the age of academic specialization, Mumford deliberately sought the rewards of breadth rather than depth, thus enabling himself to preach the same message from several soapboxes. By doing so, he ran the risk of being pegged as a dilettante. Mumford grew up in New York City and attended a high school emphasizing a technical curriculum. Enamored of the promise of technology, he was conversant in the language of his schoolmates. In a passage that shows Mumford's usual smugness and long-windedness, he recalled that he wanted to become "an engineer, and the most progressive kind of engineer, too—an electrical engineer. . . . This early acceptance of—indeed high excitement over—the doctrine of progress has given me a vivid understanding of my 'progressive' power-infatuated contemporaries."[1]

Mumford served as a radioman in World War I, never seeing combat, and clung to what he later called "an unjustifiable faith in the progressive powers of 'science' and 'democracy' to bring about a happy ending." He developed the social consciousness of a cultural and political radical after the war. At the *Dial*, he worked as an editorial assistant, met his wife, Sophia, and fell under the mystic spell of Thorstein Veblen, who "never quite lost his hold on me," Mumford remembered. In the 1920s he examined the American literary renaissance in *The Golden Day* and, following the lead of Randolph Bourne, attacked John Dewey as a soulless instrumentalist. Several tempting offers of employment came in 1927, including one for a position as an editorial writer under Walter Lippmann at the *New York World*. Mumford decided that he valued his freedom of inquiry more highly than the potential gains in wealth, so instead he began *Technics* after completing a book on Herman Melville.[2]

Judging from Mumford's correspondence of 1933, his opinion of the book, never pessimistic, rocketed higher and higher. In January he wrote Van Wyck Brooks that while he was "having fun" with the book, he was "ashamed of its Aristotelian pretentiousness." He was not yet sure if "it is either very good or quite empty." As of March, little doubt remained; the book was writing the author. "I am," he wrote, "writing steadily, firmly, relentlessly, crushing ahead as slowly as a glacier." Work was progressing "slowly but with the feeling that nothing can stop me, and that at the end I shall have a very powerful and important book to show for it." Mixed emotions surfaced in some June letters. Mumford admitted to Catherine Krause Bauer that while the book had become "gigantic," its flaws had enlarged proportionately, the way irregularities on a balloon's surface grow when it is inflated. Less than three weeks later, however, Mumford wrote Brooks with even more braggadocio. His summer, he said, was occupied by "the evil and the glory of writing" *Technics and Civilization*, which he called "a book to out-Bentham the Benthamites, to out-Marx the Marxists, and in general, to put almost anybody and everybody who has written about the machine or modern industrialism or the promise of the future into his or her place." The romance between Mumford and Bauer may account for the variation in tone; he may have felt vulnerable with her while strutting for Brooks. Either way, his opinion of his own worth was, as usual, vain.[3]

The response to Mumford's book on its publication apparently justified his faith in it. Excited reviewers repeatedly used the words *masterpiece, brilliant*, and *magnificent*, although the risks Mumford took opened him to chastisement for superficiality, inconsistency, and overwriting. Few commentators missed Mumford's central mission: to issue a manifesto for a new political-cultural order. One critic noted the elusive quality of the book's identity: "Too diffuse to be science, too concrete to be philosophy, it is difficult to fit into any category. Might perhaps be classed as one of the prophetic books." In the same genre as Veblen and Bellamy, Mumford addressed a Depression audience ready to contemplate fundamental social change.[4]

In his attempt to outline a call for cultural regeneration rather than to write a history of technology, Mumford focused on the machine because, like Veblen, he understood humanity's tools to determine its social and cultural arrangements. The machine, in turn, embodied the values of the culture that produced it. Thus, the steam-coal-steel technology Mumford dubbed "paleotechnic" continued to remind the Western world of the mining—pillage—and warfare of its origins. Modern capitalist society, in this view, tends toward regimentation and destruction of the natural world. As a result, the army represents the "ideal form" of mechanistic political arrangements, so that "war is not only, as it has been called, the health of the State: it is the health of the machine, too." Mumford's argument hinged on his organic conception of political life. Rather than follow the logic of academic specialization, he instead insisted on the interconnectedness of a society's values, economy, technology, and politics. The split of political economy into smaller specialties hindered the sort of comprehensive understanding necessary to move forward from the paleotechnic era to the neotechnic future.[5]

Blaming the machine, capitalism, or political arrangements for social problems, without seeing them as mutually reinforcing institutions, would get the critic or visionary nowhere. "Thanks to capitalism," for instance, "the machine has been over-worked, over-enlarged, over-exploited because of the possibility of making money out of it." Politics, economics, and industry, then, had to undergo synchronized adjustment. This transition required new ways of knowing. While science had provided humanity with the capacity to understand and control the natural order, to presume objectivity in the study of human institutions was dangerous folly. Those who conflated natural science with social science limited reality to data that fit their primitive models: "What the physical sciences call the world is not the total object of common human experience: it is just those aspects of this experience that lend themselves to accurate factual observation and to generalized statements." Like Frank Knight, Mumford worried about including "the spectator and experimenter in the final picture."[6]

The alternative, according to Mumford, was to realize one's own origins in the culture's past, to acknowledge historicism rather than try to conquer or escape it. In this perspective, "one knows life, not as a fact in the raw, but only as one is conscious of human society and uses the tools and instruments society has developed through history—words, symbols, grammar, logic." Even what appeared to be the "most abstract knowledge, the most impersonal method" resulted from "this world of socially ordered values." Unfortunately for humanity, the flawed objective mode of understanding tended to reify the machine. Itself the product of much objective knowledge, technology nevertheless originated with value choices. These choices, invisible from the scientific viewpoint, went overlooked as the machine took on an autonomous identity of its own, adrift from its surroundings. Machines, in Mumford's words, "have seemed to have a reality and an independent existence apart from the user." This alienation—though Mumford used other words for it—led industrial society to institutionalize an unnecessary and dehumanizing technological autonomy: "The machine has undergone a perversion: instead of being utilized as an instrument of life, it has tended to become an absolute." Here Mumford maintained his critique of Deweyan pragmatism: tools, he argued, need to be put to culturally enriching uses.[7]

To make the machine an instrument of human purpose, it must be understood in human terms. Because life is an organic whole, to comprehend the machine one must dissect the tissues of connection in which it is created and used. In Mumford's words, "No matter how completely technics relies upon the objective procedures of the sciences, it does not form an independent system, like the universe: it exists as an element in human culture and promises well or ill as the social groups that exploit it promise well or ill." No unrepentant Luddite, Mumford was admittedly fascinated by technology, which had much to teach modern cultures: "Until we have absorbed . . . the lessons of the mechanical realm, we cannot go further in our development toward the more richly organic, the more profoundly human." In fact, the soaring note on which Mumford concluded his book harmonized with the hum of the machine: "For however short modern science and technics have fallen of their inherent possibilities, they have taught mankind at least one lesson: Nothing is impossible." But he left cultural control of the machine—that would originate with the same attitudes that produced the machine—undefined.[8]

The future, recently brought within reach by the marvels of technics, had in Mumford's view to be built on values derived from the machine, then redefined in human terms. The task ahead was to "work out the details of a new political and social order, radically different by reason of the knowledge that is already at our command." That new society would "leave a place for irrational and instinctive and traditional elements in society which were flouted, to their ultimate peril, by the narrow forms of rationalism that prevailed during the past century." Thus a reordered civilization required new politics, a new economy, a new machine, and a new epistemology.[9]

In Mumford's plan, work, production, and consumption would be coordinated to enhance human life. Demand for trivial goods would be rationalized, freeing consumers from a prison of junk and saving workers from degrading labor. In a society as complicated as twentieth-century America, how could such gains be attained by a democratic polity? They would not. "To achieve all these possible gains in production . . . requires the services of the geographer and the regional planner, the psychologist, the educator, the sociologist, the skilled political administrator" in addition to the current contributions of administrators and engineers. Paradoxically, then, the way to invest life with more human values is to depend on experts whose thought has been conditioned by the premises of science and technology.[10]

For all of his talk of organicism and humanism, Mumford proposed a program not radically different from those of the rational reformers, John Dewey, or the technocrats. His powerful criticisms of technological determinism notwithstanding, Mumford relied on technological absolutism and scientific empiricism when they served his purposes. In the "cultivation of the sciences a *definite hierarchy of values* must be established," he wrote in 1922. The rationalized productive economy envisioned in *Technics* would include "scientific scales of performance and material quality—so that goods will be sold on the basis of *actual value and service*." Mumford, it appears, wanted society to pay heed to human values—if they were his human values; diversity, democracy, and inconclusiveness clearly bothered him. As Casey

Fig. 11-1. Soviet dam interior from Lewis Mumford, Technics and Civilization *(New York: Harcourt, Brace, 1934). The caption commented that the scene illustrated "the calmness, cleanness, and order of the neotechnic environment." "The same qualities prevail," it continued, "in the power station or the factory as in the kitchen or the bathroom of the individual dwelling. In any one of these places one could 'eat off the floor.' Contrast with the paleotechnic environment."*

Blake has noted, for Mumford to suggest that the same technology, "invented for profit and to further military conquest, would itself spawn a neotechnic order free of those same social forces" contradicted his own description of "the symbiotic growth of a militaristic capitalism and industrial habits of technology." Identifying the technological roots of culture turned out to be easier than cutting them.[11]

Mumford argued for a rationalized political economy that would then free humanity's irrational impulses from the iron cage of capitalist machine production. "My utopia," he said repeatedly, "is actual life, here or anywhere, pushed to the limits of its actual possibilities." The absolute terms Mumford insisted on using for the state in *Technics* rendered him incapable of inventing a world transcending its foundations; the very words he inherited precluded the escape from technical rationality of which he was so confident before the book's publication. Like Ruml and Mitchell, he substituted rational planning by experts for the regrettable inefficiency of politics, hoping that method could overcome madness.[12]

WALTER LIPPMANN AND *THE GOOD SOCIETY*

Walter Lippmann's long reevaluation of rational reform that began in World War I reached its pinnacle in 1938, when he published *The Good Society*. Doubts about efficiency, about social science, and about managerialism that had been appearing in his columns or in passing references in books and letters jelled in a diatribe against the aspirations of the rational reformers. What he offered instead—a traditional liberalism of widely dispersed property holding, of the rule of law, and of what he defined as freedom—was far less systematic than the sustained critique that dominates the book. As he had done so often, Lippmann stirred up widespread debate, and planners, rationalizers, and social engineers had to contend with the opposition and second thoughts *The Good Society* engendered.

During the 1920s Lippmann lost what little confidence he had in the capacity of the masses to think socially in scientific and rational terms. In part because of poor information and the symbolic ways in which he saw news outlets presenting complex technicalities, he steadily reversed his orientation. The rationality of *Drift and Mastery* gave way in 1929 to a call for morals rather than techniques, but even then the possibility of planning and of expert guidance (not domination) of the masses still looked reasonable. "Statesmanship," he wrote in *A Preface to Morals*, "consists in giving the people not what they want but what they will learn to want." Such a task required "an objective and discerning knowledge of the facts, and a high and imperturbable disinterestedness." In 1933, he spoke of the need to create an "ordered society," one where Americans shared "a great common purpose, disciplined to act together, educated to understand and respect superior knowledge, ready and eager to follow and to honor the leadership of our best men." Americans, he said, lived in "an age when conscious, deliberate direction of human affairs is necessary and unavoidable." But the coming of the New Deal, an administration that pointedly asked little advice of Lippmann the counselor to the mighty, forced him to oppose Roosevelt's extensions of government.[13]

In *The Method of Freedom* (1934), Lippmann began to take systematic steps in the direction of the modified libertarianism of *The Good Society* but continued to hold open the possibility of a managed (not directed) economy,

Fig. 11-2. Walter Lippmann (Yale University Archives)

which he called free collectivism. He wondered how government could legitimately oppose the popular (or mob) will and still remain democratic. While this sort of move smacks of sophistry, given his own statement of only five years before about statesmanship's giving people what they will learn to want, Lippmann did raise valid technical challenges to centralized control.

Planned economies tended to work, he argued from historical evidence, only to alleviate scarcity; to distribute abundance, markets had to provide the basic facts of consumer demand. The Soviet planners could enjoy apparent success only because of the stringent restrictions on freedom and because of the primitive state of the collective productive capacity. Lippmann had also questioned the so-called spirit of science that legitimated the planners. "Thin, dry rationalist" social scientists and democratic collectives with scientific aspirations both were found lacking. By 1935, then, not only did social science lack an adequate animating spirit, but Lippmann saw the sheer data-gathering problems as insurmountable in a planned society.[14]

Just before the publication of *The Good Society*, Lippmann had worked out his argument in columns and in letters. He pointed out the disturbing tendency among NRPB and NESPA members to stretch the term *planning*, which strictly speaking denoted "a centrally directed economy," to "loosely mean foresight and coordinated action and prudent anticipation." He also benefited from developments in European thought, specifically the writings of Ludwig von Mises and Friedrich von Hayek. Though he declared, "I had grasped the logical incompatibility of planning within the democratic system," even before reading their work, Lippmann sounded a great deal more fawning in his letter to Hayek himself. He had studied Hayek's book "very carefully and been deeply influenced by it." Without his and von Mises's books, "I could not have developed the argument" in *The Good Society*. Hayek responded to his fellow prophet, convinced that "the whole trend towards planning is an effect of a misunderstanding of 'scientific' method and a result of an exuberance about the power of human reason caused by the scientific progress of the last hundred years." A decade later Michael Oakeshott would call this link between modern liberalism and technological innovation "political rationalism."[15]

In *The Good Society*, Lippmann mounted a frontal assault on the school of thought headed by Chase, Soule, and Mumford. He discredited most of the assumptions of social engineering even while failing to propose a plausible replacement for either technocratic liberalism or plutocratic laissez-faire. The presence of European ideologies of domination provided him with an irresistible smear tactic, and Lippmann frequently tied collectivism to fascism. If advocates of social planning genuinely sought "a social order in harmony with the genius of the scientific method and of the modern economy of production," he warned them away from hardened bureaucratic structures that would encumber the "flexible, experimental, adjustable" environment that fostered innovation. That environment, Lippmann and Hayek argued, had to be competitive. To those who pointed to the abuses of free competition by monopolies and oligopolies, Lippmann responded with a confidence in litigation to ensure the proverbial level playing field.[16]

Lippmann insisted that the problems confronting those who sought to achieve social control were more complex than even the most skeptical social scientists admitted. "It is not merely that we do not have to-day enough factual knowledge of the social order, enough statistics, censuses, reports," he said in rebuttal to Wesley Mitchell and his followers. "The difficulty is deeper than that," for humanity lacked "the indispensable logical equipment—the knowledge of the grammar and the syntax of society as a whole." In the absence of such superhuman capacities, the planners would be forced to continue to expand their always imperfect control over society. Despite the claims to the contrary by Chase and others, Lippmann contended, "there is nothing in the collectivist principle which marks any stopping place short of the totalitarian state." He gave an example of the magnitude of the planners' challenge, no doubt influenced by the unsoundness of the NRA. Pointing to various commodities, he asked how the planners, without price data provided by the market, could account for buyer preference within a category—Fords versus Chevrolets—much less across categories—new cars versus new houses. The only way to get the data to fit any humanly possible model was through coercion: price controls, which had already failed, or rationing at the least. In this section Lippmann was most persuasive, but equating attempts to master social complexity with tyranny stretched things unnecessarily.[17]

In the book's conclusion, Lippmann displayed his considerable rhetorical abilities like a peacock. After alluding to Aristotle's discussion of slavery, he asked, "Are not Mr. Chase's regimented citizens mere 'living instruments' [slaves] of his glorified technicians? And as such, because they are less than men, have they not been stripped of their defenses against oppression?" Once again, Lippmann was less than fair, for Chase occupied no official position and had oppressed no citizens. Lippmann's sweeping and soaring prose, when it turned to the free market—which he equated with a free society—made workers in the age of Flint headbashing sound like kings, and consumers endangered by patent medicines seem like emperors: "It is the inviolability of all individuals which determines the social obligations of each individual." Citizens qua citizens, in such a view, were sufficiently free so as to make their unfavorable market position acceptable. In the enthusiasm—intensified by the developments in Italy, Germany, and the Soviet

Union—to decry authoritarianism, Lippmann overwrought his defense of a mythically "free" America. The dissection of the social engineering ideology, however, was often astute and damning. He saw hubris of a very real sort for what it was, and the response by the faithful constituted a significant rearguard action. Social engineering, within and outside of social science circles, had begun to recede by 1938, its promise either unfulfilled or shown to be illusory.[18]

Reviewers of the book quickly saw its problems. One remarked that if the collectivists, as Lippmann insisted, had to be supermen, "by the same token the Good Society is only for angels." Ralph Barton Perry wondered how Lippmann could look to free markets, where buyers and sellers by definition were motivated by selfishness, while also invoking an ethical norm among free men that would be crushed by planning. In the *Nation*, Max Lerner called Lippmann's liberalism "the intellectual garment of capitalist power" insofar as it neglected the reality of private coercion in the lives of Americans. (Lerner's own work resolutely defended centralized planning from a left-wing perspective: "Democratic planning may be defined as the technical coordination, by disinterested experts, of consumption, production, investment, trade, and income distribution in accordance with social objectives set by bodies representative of the majority.") Merriam and Mumford contended in separate reviews that planning was possible under democratic standards of oversight and review, implying that Lippmann had defended the class of which he had become a part. In sum, even some of those who doubted humanity's capacity to plan felt Lippmann had overplayed his praise of the common law as a tool for social control, his Hayekian linkage of planning with totalitarianism, and his attribution of benevolent motives to capitalist economic actors.[19]

A thoroughgoing denunciation of Lippmann's book, and career, appeared not in a review but in one of Charles Beard's letters to Texas congressman Maury Maverick, the leading planning advocate in the House. Beard cynically lauded Lippmann's ability to "make as clear as sunlight to tired business men and fat dowagers who read the *Tribune* things that are clear only to Almighty God." Lippmann was "a befuddled man now wining and dining with the boys of the main chance. He loves their praise and their fleshpots." The moral pretension present most odiously in the conclusion to *The Good Society* particularly bothered Beard: "Over all his ratiocinations he sprinkles the odor of a sickly humanism as thick as the smell of magnolia blossoms which apologists of the good old days in the South spread over the sweat of slave gangs." An apologist for the plutocracy, Lippmann was a facile alarmist who "does things to [Beard alluded to Jonathan Swift's original verb, in reference to King James I: "beshit"] himself when there is no danger." Beard's vitriol could have resulted from envy, for as a freelance intellectual he lacked plutocratic accouterments, or from the mischievous talent he had with words. In any case, their own practical shortcomings and the shift in both Roosevelt's and public opinion away from the ideology of technocratic social control put Beard, Merriam, Lerner, Mumford, and the other advocates of various degrees of centralized government rationalization on the defensive.[20]

Among the defenders of social technology, John Dewey characteristically took *The Good Society*'s criticisms seriously even as he saw its limited value as a guide to action. He noted the book's tendency to "give encouragement and practical support to reactionaries," while finding Lippmann to live, ultimately, in a "vacuum"—the only place such a perfect market could exist. Even so, Dewey found himself "completely in agreement with Mr. Lippmann's indictment of the authoritarian state" as a means for realizing authentic social goals. Like others, Dewey had reconfigured his reliance on scientific methods; he concluded that finding the tools—a detail Lippmann avoided—by which a "free society of free human beings" could be reached constituted the "central problem, intellectually and practically, of genuine liberalism today." The means that had held so much promise were by 1939, like so much else, chimerical and fallible.[21]

AMERICAN SOCIAL SCIENCE AFTER 1935

With the decline of support from the Rockefeller Foundation for the Ruml-Day program of social science and social engineering, the momentum that had developed in the pre-1932 period shifted direction. While a systematic investigation of social science agenda setting in the 1930s is beyond our present scope, some general categories of reconsideration can be discerned. A number of sources suggest that a transition occurred as the leaders of the 1920s—Ruml, Day, Mitchell, and Merriam—worked in other areas in the late 1930s. The Rockefeller administrative staff, meanwhile, asked new questions and reopened old ones, especially after 1935. The SSRC organized a concerted stock taking in 1937 to chart its past and debate its future, and in 1939 Robert Lynd issued a strong challenge to his colleagues from within the profession. As these investigations tested the premises of social engineering, differing conclusions emerged. Depending on the document, it was either the dawn or the twilight of the program to invigorate social reform with science.

. . . .

Representative of the uncritical dogmatists who remained committed to social engineering, Alexander Goldenweiser and Howard Odum made declarations that echoed earlier arguments but by 1939 found less support in professional social science. One of the prime movers behind the *Encyclopaedia of the Social Sciences*, Goldenweiser continued to preach an American Comteanism that placed the mantle of progress squarely on the shoulders of the social scientist. As he wrote in 1936, the "social disciplines" looked forward to "the enhancement of rational social action" along the lines of natural science. After discoveries were made, the method of social engineering centered on education. "In a democracy," he continued, "the first task is to mould the public mind—should it require moulding—into an attitude favorable not to a particular scheme, but to schemes of change in general." With Adolph Hitler doing just that in Germany, however, many fewer social scientists endorsed such a vague mandate.[22]

Odum, an evangelist for scientific social planning whose ardor had not diminished, kept his faith in the rational reform model, perhaps because he had met with more success in the regional planning movement than the national planners did. Yet he also maintained his national focus. American progress could only be maintained, he argued vacantly, "through the matching of technology with more and more comprehensive and effective social technology—social study, social invention, social planning, social action— both symbol and actuality of a new social constitution." He continued to endorse social engineering after many other enthusiasts had pulled back; in his view, social technology "after all is nothing more nor less than technical and practical ways of attaining social ends." The perceived value neutrality of the scientific method also provided Odum with a selling point for a textbook. As he told his publisher, "The book is an epitomy [*sic*] of the new liberalism in America, safeguarding students against emotional movements of radicalism or reaction." Odum asserted that the book spoke to students who needed to understand society "without, on the one hand, feeling the urge to overthrow something or, on the other, to become discouraged." He tied the apolitical stance of science with sociologists' aspirations for political power and tried to market the result to students.[23]

That text, *American Social Problems* (1939), was from start to finish a sustained tocsin for scientific planning as a salvation of democracy. Odum brought together, perhaps more clearly than in the movement's prime, the elements of social engineering: the presumption of apolitical expertise, the self-interest of the planner class, and the unquestioned analogy between the social and the natural environment. He opposed, early in the book, the "scientific-liberal" outlook to a number of less savory alternatives, making his program the sole logical alternative. "The dogmatic-conservative, the emotional-radical, or agnostic-objective [viewpoints], on the assumptions of tradition, philosophy, ethics, or 'pure' science, deny the effectiveness of social science and social planning," he wrote. But even after the challenges raised by Hayek, Lippmann, and others, Odum refused to allow that many critics could both identify themselves as liberals and oppose social engineering not because it lacked effectiveness, but because it was not appropriately democratic or republican. The book, meanwhile, baldly sought to elevate the prestige of Odum and his fellows, saying that "it would be well if the common man were to hold the leaders of social science in greater respect than he has done in the past." He assumed that his goal—"the rational regimentation of irrational society"—was, or should be, universally held, so the possibility of credible opposition to his plan was simply not entertained. By 1939 anthropologists following the lead of Franz Boas had begun to argue for cultural relativism in contrast to the assumed absolutism of models like Odum's. Fewer intellectuals called for anything as simplistic as "social invention and technology for the mastery of the new social frontiers as the old technology mastered the physical frontiers" or for better "technicways," as he called them, but Odum continued to put stock in an idea and a language whose value had plummeted and that paid fewer dividends every year.[24]

....

As social science came to what one historian called "a complete impasse" with regard to goals and methods by the end of the 1930s, the confident assertions of Goldenweiser and Odum had become rare. Within the leadership of the social sciences, substantial rethinking was under way. In his internal report on the history and prospects of the SSRC, the political scientist Louis Wirth contributed to a revision of the group's governing ideology by pointing up some of the implicit reliance it had on the engineering model. He began by dismissing the analogy between social science and natural science, in part because it appeared to guarantee certainty where none existed. Wirth saw "a striking difference between knowledge about social affairs as distinguished from knowledge of physical things," noting in particular a difference in "the control they give man in solving his problems." He also found the mythologized objectivity to be "more difficult to attain" in the social sciences. Finally, "Discussion about human affairs, moreover, must proceed by means of a language the terms of which do not refer back to strictly denotable objects in the perceivable world. This introduces ambiguity and results in unstable frames of reference. Despite strenuous effort to achieve consensus on essential terms, it frequently happens that different investigators talk past rather than to each other." If science could not bring unanimity among social scientists, what hope existed for agreement among a wider, less expert public?[25]

In addition to acknowledging, but not objecting to, the SSRC's role as a money laundering enterprise—"as a 'purifying' agency for research funds" —Wirth attempted to break the connection that Edmund E. Day and others had established between investigation and application. "It is impossible," the report argued, "for the social scientist, if he has due regard for the nature and limitations of his knowledge and its relation to social action, to set himself up as competent to prescribe what society should do to solve its problems." Wirth did not deny the social scientist's role as a citizen but made it clear that the program of social engineering included in the SSRC mandate, in particular through the Committee on Problems and Policy, had itself to be purified. For the SSRC to retain credibility in an intellectual milieu that valued precision, abstraction, and distance from rather than proximity to the centers of political power, inquiry, not application, had to be reemphasized. The logical positivists and the social scientists who would be responsible for the boom in behavioral sciences after World War II put a premium on new readings of "scientific" in their sociology of knowledge. The result struck yet another blow against the empire of prewar scientism.[26]

At the ten-year anniversary of the University of Chicago social science research building in 1939, more evidence could be discovered of a significant challenge to the social engineering model. Discussions of the quest for precision, factor analysis, and other newly sophisticated research methodologies coexisted with calls for effective teachers and, from William Ogburn, a reiteration of the need for separating research from application. The most confrontational episode, however, occurred at the opening address, given by Robert Maynard Hutchins, the university's president. Even though he had been a viable candidate for a position in the NRA, and thus not antagonistic to the idea of rational control of social forces, he had no time for the pretensions that marked social engineering's claims.

Hutchins got quickly to the kernel of the matter: "Now, instead of cherishing the hope that . . . we could rapidly achieve a better understanding of our society, we must ask ourselves how long there is going to be any society to understand." He directly attacked the pretexts upon which the building was founded—objectivity, expertise, and the emulation of natural science. An undoubtedly seething audience was told that "merely calling ourselves 'scientists' isn't going to help." Instead, doing so more likely would lead investigators to assume that their "object is to improve the material conditions of existence." The problem, he reminded his guests, was that "now the great problems of our time are not material." Instead, they were "moral, intellectual, and spiritual," and technique offered a poor substitute, in this understanding, for values. Hutchins's diatribe foreshadowed a whole genre of wartime writing concerned with mounting a moral defense of America, in part to distinguish it from the scientific monstrosities of the Axis powers.[27]

. . . .

Within the Rockefeller social science bureaucracy, similar second thoughts, rationalizations, and redefinitions were also under way. The staff member John Vansickle, for example, told Raymond Fosdick in 1937 that the social scientist must work "within limits that are far narrower than those prevailing in the exact and experimental sciences" to make educated predictions, not to declare truths. "Final decisions," on the other hand, "depend on social attitudes and social values and these are not matters of scientific determination." Simon Kuznets, an economist at the NBER, wrote Joseph Willets, a Rockefeller Foundation board member and dean of the Wharton School at the University of Pennsylvania, on the same topic. "It is not the function of social scientists as scientists," he said, "to apply this knowledge to certain specific situations and thus become, as you put it, assistant social engineers." Kuznets insisted on "a clear distinction between the functions of the scientists and those of the appliers," a distinction frequently blurred during the 1920s, when "objective" scientists sought both credibility and influence. Shifts in funding patterns and the presence of many similar letters suggest that such opinions had been solicited to help the foundation change direction.[28]

At all levels of the philanthropic social science network, fundamental questions emerged. Even Day himself, after leaving the Rockefeller office, asked E. B. Wilson, the former head of the SSRC, "what is it that we have in mind when we so glibly refer to scientific work in the social field?" He had no real answers, and until some could be found, "we shall have to quit selling our wares as we have been doing." Day's successors at the General Education Board, in the preparation of a conference on social science education, also raised a question that for so many years had been assumed to have been self-evident: What is the end to which social scientific techniques are put? One indication that times had changed came when the conference's working group was told to differentiate between "scientific problems [that] are problems of knowledge" and "practical problems [that] are problems of action."[29]

Because the process of social triage required a normative context, "a statement of the values basic to a liberal-democratic society" needed to be explicated. That list constituted a rough baseline; in the words of the organizers, "situations which diverge from, undermine, or block the realization of these values constitute social problems." Even so, the spongy definition of liberal-democratic society simply incorporated a new understanding of social engineering. The list of liberal-democratic values illustrates this imprecision.

1. Material well-being of the entire population, including
   a. physical health
   b. adequate supply of material goods and services
2. Psychic security
3. Opportunity to compete for the possession of necessarily scarce values on the basis of competence in the performance of tasks for which those scarce values are rewards
4. Opportunity of the populace to participate in the determination of policies and measures affecting them
5. Responsibility of elected and appointed officials to the populace or its elected or otherwise designated agents
6. The maintenance of civil rights, including
   a. freedom of speech
   b. freedom of press and publication
   c. freedom of association
   d. freedom of worship
7. Opportunity for personal self-development so as to enable the individual to realize those of his potentialities which are not socially deleterious, including
   a. freedom of occupational choice
   b. freedom of consumption choice.

The very bureaucratization of civic and personal life embodied in such a document reveals how social scientists' problem solving had evolved. Comte's, Ward's, and Small's goal of social control—not only over criminals and deviants, but over the masses—came to be realized in ever more rarefied and less public forms in such entities as the conference. Even as the organizers simultaneously questioned and built on the work of their predecessors, the footings of the problem-solving edifice were sunk in the bedrock vision of science as solution.[30]

. . . .

The most significant prewar internal reformulation of the social engineering ideology occurred in 1939 with the publication of Robert Lynd's *Knowledge for What?* Because Lynd (1892–1970) was unique in that he had both received patronage directly from John D. Rockefeller, Jr., for his early religious and sociological studies and worked as the executive secretary of the SSRC, he knew the social scientific world of which he wrote from several perspectives. He used his book to challenge his profession, raise value questions to the level of discussion, and take some shots at his Columbia colleague Wesley Mitchell. When all the disruption settled down, though, Lynd had left undisturbed the fundamental precepts of the social engineering program.

After decrying the "sinister partial impotence into which progress has led us, despite the fact that ours is physically the most superbly endowed culture on earth," Lynd put forth a vision of an activist social science. Such a venture would proceed cautiously, self-aware of the "halo of adequacy which the term 'science'" lends it. Nevertheless, Lynd called upon social scientists to abandon, at least in part, the tendency to hug empirical studies of phenomena gone by. Such an attitude, he argued, allowed investigators to perpetuate the status quo—"We social scientists tend to begin by accepting our contemporary institutions as *the* datum of social science." These institutions conveniently comprise a "system" operating according to "laws," which turn out to be the governing ideas of the present. Instead, Lynd argued, science will not discover order within human groups unless order is "*built into it by science.*" But to construct a rational future instead of explaining away the present, scientists had to stop piling up data and begin asking value questions: "What do we human beings want this particular institution-complex to do for us, what is the most direct way to do it, and what do we need to know in order to do it?"[31]

Lynd then subjected each of the main social sciences to a brief critique of current methods and orientations. He singled out economics for a particularly pointed attack on the subject of assumptions. Research work like that produced by Mitchell's NBER "tacitly assumes that private, competitive business enterprise, motivated by the desire for profit, is the way for a cul-

*Fig. 11-3. Robert Lynd (*Saturday Review*)*

ture to utilize its technical skill to supply its people with needed goods,"
Lynd argued. Similarly, he thought political science neglected the central
question: how to reconcile democracy with the inherent inequality of the
species. In both cases, and in other disciplines as well, "a science jeopar-
dizes its status as science when it operates uncritically within the grooves of
folk assumptions." A predictable proposal, given Lynd's SSRC background,

was the softening of specialization in favor of interdisciplinary problem-oriented task groups focusing on topics like labor, child development, and the family. The social sciences collectively, meanwhile, needed to collaborate on the development of a working social theory to replace the dead rationalism and abstruse specialization of philosophy. Such a coordinated complex of investigators would, Lynd reasoned, be up to the task of an activist social science: the design and construction of a rational society.[32]

Lynd chastised social scientists who, assuming the posture of disinterestedness, let value questions in "through the back door." Here Mitchell came under direct criticism. "One cannot assume that the meanings of 'facts' are always clear or unequivocal," Lynd wrote. "Somebody is going to interpret what the situation means." That somebody could be a business executive, a merchant, a labor leader, or an advertising writer, none of whom, in this formulation, could see society from the constructive and global perspective of the social scientist. The task of these investigators, for Lynd, "is to find out ever more clearly what these things are that human beings persist in wanting, and how these things can be built into culture." When those desires turned out to be "ambivalent," and if humanity "is but sporadically rational and intelligent," social engineering and social control become the proper mission of the social scientist. His or her task is to discover "what forms of culturally-structured learned behavior can maximize opportunities for rational behavior where it appears to be essential for human well-being." How far such apparently behaviorist conditioning would reach was left unclear, but the prospect remains frightening in any phrasing. He also proposed what might be called cultural engineering; social science should "provide opportunity for expression of his deep emotional spontaneities where those, too, are important." The desire to remake the world, borrowed from the mastery of the natural environment, continued to condition the aspirations of social scientists even when they tried to reorient their field.[33]

In his final chapter, Lynd enumerated a series of "outrageous hypotheses" that he thought challenged the sanctity of democracy, capitalism, religion, and other givens of American life. The crisis mentality of the late 1930s clearly informs the book; as Lynd wrote, "it is an exceedingly narrow and hazardous path we social scientists must here explore," with fascism casting a lengthening shadow. Democratic social engineering, not far removed from what Dewey had argued for, remained at the center of the *Knowledge for What?* universe. The faith in rationality, in the efficacy of applying rational plans, and in the social scientist as a benevolent administrator and planner led Lynd to build the hopes of the discipline higher rather than to deflate the pretensions of the apologists.[34]

Response to the book, swift and vigorous, varied according to the scientism of the reviewer. Charles Beard approvingly congratulated Lynd on his recognition that value questions informed even the most piously "objective" student of society. Stuart Chase noted his assent. Max Lerner "salute[d] also the newer spirit of an unashamed instrumentalism in social science" before expanding on Lynd's argument. Given that current work in the field was run by foundation administrators "whose thought-processes have already been shaped in the image of the power formations of the day," an alternative was necessary. "If our democracy were worth its salt, it would create a Research and Planning Commission to take over the functions of the foundations," and on such a body, Lerner concluded, should "be men like Robert Lynd." In designs like these, Chase's class of philosopher-engineers gave way to a breed of philosopher-sociologists.[35]

Opposition to the book came from inside and outside the social science community. The outsiders, most notably the historian Crane Brinton, attacked the paternalism that would lead uplifters like Lynd to put social scientists in control before they bothered to get the basic data right. "Most sociologists," Brinton wrote, "are so interested in trying to devise ways of improving man's behavior that they neglect the less noble but more useful task of observing that behavior." Even though the criticism might be slightly unfair, given Lynd's record of direct observation, Brinton still pinned down the central dilemma, also revealed in *Middletown*. Sociologists could not trust "the ordinary American, the man who tunes in on Father Coughlin, reads the *Saturday Evening Post* or *True Story*, throws orange peels and cigarette stubs out of his Ford," yet they persisted in "'planning' all sorts of nice things for him." When Lynd called for expertise in the administration of the country and the maintenance of democratic institutions, Brinton argued, something rang false. Alluding to the book's use of Auden's remark about lecturing on navigation while the ship is going down, Brinton corrected Lynd: "Navigation owes a lot to the 'pure' sciences of astronomy, physics, and meteorology, but it remains a practical skill among men trained for a life of action. Ships aren't run by scholars from laboratories. Sociology might be more useful if sociologists ceased to try to run the ship of state from their studies, and contented themselves with the job of making sociology a respectable science."[36]

Trying to increase the "scientific" accuracy of social scientific data occupied Mitchell throughout the latter half of his life, so that the bruises Lynd inflicted on his reputation and outlook were painful. The old dilemma of advocacy and objectivity replayed itself in the strained relations between the two men after the book appeared. Lynd, who told a class at Columbia that "'knowledge for its own sake' is not science," held that "the only justification (other than a personal one of, e.g., a hobby) for existence of any social science is that it is working as a tool of man in solving man's problems." Mitchell obviously saw things differently. After the episode, Lynd offered a more charitable defense of his attack on fact gathering: "I've always felt that there are two persons in you: (1) the man of 'The Backward Art—' and 'Human Behavior and Economics,' (2) the man who committed himself to empiricism in part due to your experience in Washington 25 years ago. A lot of us younger men, sore beset by current problems, would like to see the earlier Wesley Mitchell come to the fore again." That earlier Wesley Mitchell, though, however much he believed in scientific social progress, also had aged and been made wiser by experience. The strain between the two sides of his personality showed through most evidently in the 1930s.[37]

In lectures, informal talks, and writing, Mitchell tried to juggle the need for control over complicated social forces, the presence of a democratic citizenry, and the problems of a science with a limited supply of "solid" facts. As early as 1934, he asked how planners could understand "themselves fit to tell their fellow citizens what ought to be done." The possibility of scientific unanimity, perhaps through education, failed to impress him, for "as soon as we begin discussing what we mean by social welfare or what we think is for the best, we are back at the starting point": political discussion, contention, and relations of power. While the World War I planning experience, when the goals were clearer as politics receded in importance, had shown the engineers to be the people best prepared to meet the emergencies of production and distribution, even they "were in very large cases more or less disabled from meeting the situation because they so often forgot that human beings are most variable and intractable materials." The resulting understanding—that social scientific knowledge had limits, that democracy implies politics, and that goals could only rarely steer techniques to proper use—made Mitchell hesitant to prescribe remedies, no matter how pressing society's problems.[38]

Lynd's charges deeply hurt Mitchell, who had been pondering these issues and chiseling out his own reconciliation for decades; "metaphorically speaking," he told the Columbia Economics Club, "Lynd and I belong to the same church; but in his eyes I am a backslider." Later he said, even more poignantly, that "Lynd makes me feel like a fallen angel." In the discussion, Mitchell took *Knowledge for What?* seriously yet trusted his own experience to demonstrate the inadvisability of handing power to the social experts. The economist himself, as opposed to the caricature in the book, did not oppose social scientists speaking out on issues—if they left their credentials at the door. "But when you do give advice," he cautioned, "don't pretend that you are speaking as a scientific man—unless you *know*." The inadequacy of social and economic data in his own professional life made Mitchell wary of Lynd's proposals, in part because investigators had proven themselves especially incapable of predicting the negative effects of the solutions they proposed for current problems; social cures often harmed more patients than the disease had.[39]

Mitchell concluded by telling the listeners a cautionary tale that Lynd, after so long in academic life, may not have understood. The hopes, frustrations, and second thoughts—about economics, about America—after a career in the forefront of rational reform must have moved his audience: "Let anyone who thinks himself equipped to reconstruct American culture first try to settle some issue so small that he can learn what needs to be known. . . . For example, let him study the problem of stopping the pollution of streams by industrial wastes. The experience is likely to leave him a sadder and a wiser man." Mitchell's ceaseless search for more accurate figures and a sounder base of knowledge enabled academic economics to grow in prestige, but his real goal, that of a better world, remained ambiguously unattained. In 1940 he wrote that while "social engineering is indeed much needed," it was premature to turn the NBER in that direction, "for the folk who give their strength to economic reforms do not have such well-tested knowledge of economic processes as engineers possess of physical processes." Science, or their version of it, turned out to have less potential for social improvement than many reformers had envisioned.[40]

. . . .

Never satisfied with any one route to social change, Beardsley Ruml served as a dean at the University of Chicago and as the treasurer of Macy's after leaving the LSRM. Following his moment of fame in World War II, Ruml continued to reflect on methods for social improvement. Late in his life he expressed discouragement to his close friend Charles Merriam that he had not found the right vehicle for his vision: "As I reflect on the history of the last thirty years in social science, I am inclined to think that one reason so little progress has been made is that the boys were more interested in getting their

money than they were in the development of their subject matter on a fundamental basis." Because so much good work in social engineering occurred
outside the universities, "there must be something deeply wrong [with them]
as a setting for social research." But Ruml's belief in applied science as a
model for reform had yet to wane in 1951. He concluded his letter to Merriam by reporting that "I am leaving with Stuart Chase for Puerto Rico [where
Tugwell had been governor in the early 1940s] on Saturday. Good social
technology going on down there, but again, mostly not in the University."
Such an admission had to trouble Ruml, who had committed so many resources to the development of the academic social sciences.[41]

. . . .

Through the 1930s, Charles Merriam continued to try, with no real success,
to bring together planning, democracy, and social scientific knowledge in a
program of political rejuvenation. A particular rebuff of his outlook, which
he thought was based on misrepresentation and therefore felt all the more
frustrating, took place near the end of World War II. At a radio forum featuring Merriam, the Socialist political candidate and economics professor
Maynard Krueger, and Friedrich von Hayek, debate got particularly nasty.
Merriam's central assertion—that planning led "toward freedom, toward
emancipation, and toward the higher levels of human personality"—was to-
tally denied by Hayek, who repeated his familiar line that totalitarianism
and planning were inextricably connected. Hayek's refusal to engage Merriam at even the simplest level illustrated most painfully how far the ideas of
even an antidespotic social planner had fallen.[42]

Over the course of the broadcast, Hayek attacked each of Merriam's pillars of a planned democracy. Merriam's vehemence, in turn, remains palpa-
ble. The degree to which his concepts had hardened into concrete and had
ceased to be hypothetical, meanwhile, was pointedly demonstrated by
Hayek, who challenged them explicitly:

> Merriam: . . . The great gap in Hayek's studies, among many gaps, is
> that he does not reckon with public administration and with manage-
> ment. He regards anything that is delegated to an administrator or man-
> ager as being irrational, if I understand him correctly.
> Hayek: There are so many points [to address]. Krueger, yours is, I be-
> lieve, the most important.
> Merriam: I regard mine as most important.
> Hayek: I must begin with Krueger. . . .

[A few minutes later, the Austrian gunned another of Merriam's cherished notions down without ceremony.]

Merriam: . . . You seem to express grave doubts about the ability of a democratic society to accomplish very much. . . .

Hayek: I am saying that people like you, Merriam, are inclined to burden democracy with tasks which it cannot achieve and, therefore, are likely to destroy democracy.[43]

Hayek's vigorous embrace of the idea of competition, dismissed by the rational reformers a few decades before as a quaint but dangerously outmoded relic, confronted Merriam with the need to defend his primary assumptions anew. The premises of Merriam's brand of social engineering—that education in more "scientific" modes of civic thinking could attenuate conflict, that administrators could overhaul parts of the economy, and that scholarly objectivity could inform bureaucratic fairness—were all denied by Hayek, but those premises had been less than solid for some time. In the context of the broadcast and afterward, when Merriam said that "Hayek's trip to campus stirred up a good deal of bitterness and, on the whole, did little good," his irritation becomes more understandable when we realize that the three men wrangled for six hours the night before the broadcast. "Even so," Merriam told Ruml a few weeks later, "we got nowhere." The same might be said for Merriam's efforts to instill the spirit of planning in the American psyche.[44]

Nineteen thirty-nine was a year of anniversaries. Forty years earlier, welfare and charity workers in New York called what they did social engineering and established a journal of that name. In 1909, Herbert Croly had insisted that for America to realize its promise, rational administrators needed to be given broader control. Nineteen nineteen forced many social scientists who had mobilized the country for war to reconsider their designs for centralized control over industry. Of course, 1929 began a decade of stringent economic and political conditions that rationality could not solve. At its tenth anniversary the University of Chicago social science research building stood not as a center where definitive answers could be found but as a reminder of the Rockefeller Foundation's hopes for social engineering. More recently, 1934 had witnessed the failure of the NRA, the publication of *Technics and Civilization*, and Herbert Hoover's attacks on his successors.[1]

In 1939 defenders of social engineering retreated as challengers were emboldened by the growing understanding that irrationality, in a democracy especially, could not be contained by laws, theories, or logic. Friedrich von Hayek called for renewed appreciation of the magic of competitive markets; Reinhold Niebuhr wanted Americans to restore their respect for the concept of evil; and movies such as *The Wizard of Oz* and *Gone with the Wind* illustrated a desire to escape, through technology, the very world that technology had promised. Amid prolonged economic malaise and European militarism, more and more Americans resisted the narrow view of reality upon which social engineering was premised. Manipulations of consumptive desire, of irrationality, and of mass politics could be seen on many fronts, each refuting some premise of rational reform.

....

In their desire to sell the scientific method to Americans as a solution to any difficulty, the organizers of the 1939 New York World's Fair erected a monument to much more than they might have intended. The conscious messages the fair projected in cumbersome and artificial exhibits such as Democracity were overtaken by the more pressing needs and powerful methods involved with making a profit. Political regeneration no longer originated in the productive and inventive engineering feats that the fair ostentatiously celebrated. Rather, America and the fair existed within an ethos of salesmanship, sex, and consumption reinforced by the application of psychoanalytic research to the problem of buyer motivation. Symbolic constructs abounded: the fair was built on reclaimed land, the site of the valley of ashes in *The Great Gatsby*; General Electric demonstrated its potency in Steinmetz Hall; and, most significantly, RCA and NBC launched regular television broadcasting at the fair's opening on April 30. If one recalls that the medium had first been demonstrated with Secretary of Commerce Herbert Hoover on one end of a hookup, the cultural meaning of the second half of the twentieth century crystallizes in a fallen administrator-president, the glitz of the fair, and the politics that resulted from the marriage of the two.

An ideology holding that applied scientific rationality could solve social problems was woven into the synthetic fabric of the fair from the outset. Lewis Mumford and other professional planners helped focus the fair on the future made possible by science rather than on a celebration of the success enjoyed to date. Rationality permeated the design of every exhibit; as one observer reported, "Modern industrial 'technology' and the organization of every sphere of the life of a modern nation are illustrated in direct, concise ways in the Fair." Even the official guidebook claimed that "science is also shown as a way of solving problems." This underlying pragmatism, the testing of ideas in results, was inescapable. One visitor later recalled the fair's functionalist imperative: "What I was to understand about Tomorrow was that it Got Things Done, just the way the Borden Company got the cows milked around a giant, sterilized, udder-pumping merry-go-round called a 'rotolactor.'"[2]

Given the gravity of the world situation, the organizers of the fair took it as their morally uplifting task to teach visitors to adapt to the future. The science director of the fair proclaimed it essential for modern citizens to possess "a spirit of scientific curiosity and a willingness to apply the scientific method to social problems." H. G. Wells drove home the point that civic education in the ways of science would turn back the forces of chaos. "We are in the darkness before the dawn of a vast educational thrust," he wrote in the *New York Times* World's Fair section. "In the near tomorrow a collective human intelligence will be appearing and organizing itself in a collective human will." Even the Home Furnishings Building taught the lesson of scientific salvation:

Fig. C-1. Herbert Hoover at the first public demonstration of television, 1927 (Herbert Hoover Presidential Library)

> Time for Interest in Government
> In Community, in the Group.
> Time to Plan for our Community.
> At last Man is freed
> Freed in Time and Space.
> For What?
> . . . We may derive comfort and inspiration, however, from the fact that
> science has provided us with the means—the tools—to shape a better
> world.

In essence, William Ogburn's cultural lag idea underlay all the pretension. If the motto of the 1933 Chicago fair (Science Finds—Industry Applies—Man Conforms) no longer held, the aura of technological determinism still confronted fairgoers.[3]

Grover Whalen, the president of the fair corporation, distinguished himself both by budget-busting self-aggrandizement (with office walls made of copper) and with his careful attention to commercial interests. He told prospective exhibitors in 1937 that "of immediate benefit will be the new selling theme, the new stimulus to buying that the Fair will provide." The sanctity of science would enhance a positive environment for product placement, a goal rather distinct from the creation of a scientific democracy, his other stated objective. He told the mass assembled at the dedication of the fair's "theme center" that "this Fair, your Fair, is determined to exert a social force and to launch a needed message." To help do so, Whalen insisted in the initial stages upon a wholesome midway, in contrast to previous fairs at which peep shows had been the most profitable exhibits. His vision of a rationalized paradise was quickly punctured by the harsh reality of low attendance. Girlie shows then multiplied to the point where visitors could choose from among eleven different displays, including Dream of Venus designed by Salvador Dali and Crystal Lassies, the mirrored contribution of the industrial designer Norman Bel Geddes. At the Congress of World's Beauties, the concession featured "room for several thousand people to view the devotees of health through sunshine," while Amazons in No-Man's Land displayed "young athletic women" whose "acts and manners remain on a high plane of artistic and gymnastic achievement."[4]

The high rationality of skyscraper America failed to make the fair a moment of political reawakening. Instead, the realm of irrational desire provided the stronger message, as sexual titillation and consumptive fantasy contributed to a prophecy of the future far truer to life than the sterile geometry of the engineers-turned-shamans. As one student of the fair wrote: "Conceived as a demonstration of the triumph of enlightened social, economic, and technological engineering, it was in actuality a monument to merchandising." Such an outcome merely replayed what was happening at large. Engineers' innovative capacities still inspired wonder, but consternation and fear now entered the equation as well. The world of material goods made by the engineers became the environment in which the newer forces, advertising and marketing among them, came to the fore. Eventually tools of persuasion and selling, conveyed through the mass electronic media and appealing to urges the rational reformers hoped would go away, reconfigured American politics in ways the social engineers could only envy.[5]

. . . .

Citizens can know the abstraction called the state only through symbolic constructs. Rather than serving to represent America as a democratic republic grounded in history, the image of the state as dynamo or bridge or sky-scraper imported ahistorical notions of antidemocratic hierarchy along with the promise of efficacy. The pursuit of static but methodologically right answers ignored the processes whereby societies come to contingent and working understandings always subject to dispute. Rational reformers seized upon mechanistic metaphors for many reasons, but most neglected self-critical reflection that would reveal the limitations of this approach to social change. Logic was correct, quantification was precise, and effects resulted from discernible causes, or so these people assumed.

The pace and scope of industrial development made the engineer a ready referent and a powerful explanatory device, as simple reasoning appeared to reveal. Control of nature foretold mastery of society. Mass production promised abundance on a scale previously inconceivable, so that only rational distributive arrangements remained to be accomplished. Certainty and precision appealed to those daunted by ill-defined ennui. Having a means for determining right answers would promote consensus, as debate would become superfluous. What, then, are the consequences of such an orientation for problems of meaning as opposed to technique, for multiple frameworks of understanding, and for inherently irrational human beings? The myth of social engineering disallows statesmanship, honest disagreement, and problems that have no solutions. In a democracy, such ongoing dilemmas contribute to frustration and inefficiency, to be sure, but to solve the fundamental contradictions of the modern state simply by ignoring them is an act more metaphysical than any of the religious rationalisms against which the empiricists revolted. An artifice of supposed certainty, however, promised a secure holding place amidst buffeting winds of change.[6]

At the same time that reformers looked to an idealized version of applied science as a method for political rejuvenation, engineers and other system builders developed new technologies, of communications especially, that decisively remade the world while the technocratic reforms sputtered. The initial thrust of twentieth-century technological development tended to emphasize production and construction, so the social technicians of the 1920s and 1930s held an idealized notion of the good they could achieve by adapting the logic of making things. But by the same period, material abundance had become enough of a reality that other sciences of human behavior started to investigate human consumption of goods. Thus advertisers, who applied psychological insights to masses of individual consumers, soon were infinitely more important than the social theorists who attempted to rationalize

America. Those who sold things via new mass media successfully exploited human irrationality while social engineers worked futilely to make society logical.

That they were largely unable to do so comforts the observer only briefly. Techniques promising to end the need for ideology failed to do so. Instead, they depersonalized civic authority to the point where bureaucracies resemble the faceless despots Hoover so feared. The problems we have inherited from the engineers and the politicians appear little short of intractable. But indiscriminate condemnations of engineering often replace equally unfounded celebrations of it; America remains a largely misunderstood technological republic. In the meantime, research, development, discovery, and invention continue to redefine the physical, intellectual, and political world. Until we understand these processes more completely, citizens will continue to express frustration at having nowhere to turn—for sources of inspiration, for a language of articulate discontent, or for creative solutions.[7]

NOTES

In addition to the abbreviations found in the text, the following abbreviations are used in the notes.

For archival materials:

B   Box
DC   Document Case
F   Folder
RG   Record Group
S   Series
Sub   Subseries

AJS	*American Journal of Sociology*
APSR	*American Political Science Review*
Beard papers	Charles A. Beard papers, DePauw University Archives, Greencastle, Ind. Used by permission of Arlene Beard and Detlev Vagts.
CoP papers	A Century of Progress Records, Special Collections, University Library, University of Illinois, Chicago
Cooke papers	Morris L. Cooke papers, Franklin D. Roosevelt Library, Hyde Park, N.Y.
Day papers	Edmund E. Day papers, Rare and Manuscript Collection #3/6/8, Cornell University Library, Ithaca, N.Y.
Delano papers	Frederic Delano papers, Franklin D. Roosevelt Library, Hyde Park, N.Y.
Dennison papers	Henry S. Dennison papers, Baker Library, Harvard Business School, Boston, Mass.
ESS	*Encyclopaedia of the Social Sciences*. 15 vols. New York: Macmillan, 1929–35.
HHPL	Herbert Hoover Presidential Library, West Branch, Iowa
PS	Public Statements file
JD:LW	*John Dewey: The Later Works, 1925–1953*. 17 vols. Edited by Jo Ann Boydston. Carbondale: Southern Illinois University Press, 1981–90.
JD:MW	*John Dewey: The Middle Works, 1899–1924*. 15 vols. Edit-

	ed by Jo Ann Boydston. Carbondale: Southern Illinois University Press, 1976–83.
Johnson papers	Alvin Saunders Johnson papers, Manuscripts and Archives, Yale University Library, New Haven, Conn.
Lippmann papers	Walter Lippmann papers, Manuscripts and Archives, Yale University Library, New Haven, Conn.
Lubin papers	Isador Lubin papers, Franklin D. Roosevelt Library, Hyde Park, N.Y.
Lynd papers	Robert S. Lynd papers, Library of Congress, Washington, D.C.
Merriam papers	Charles E. Merriam papers, University of Chicago Archives
Mitchell papers	Wesley Clair Mitchell and Lucy Sprague Mitchell papers, Rare Book and Manuscript Library, Columbia University, New York, N.Y.
NR	*New Republic*
Odum papers	Howard W. Odum papers, Southern Historical Collection, University of North Carolina, Chapel Hill
Ogburn papers	William Fielding Ogburn papers, University of Chicago Archives
RAC	Rockefeller Archive Center, Pocantico Hills, N.Y.
GEB papers	General Education Board papers
Jones papers	Mark Jones papers
LSRM papers	Laura Spelman Rockefeller Memorial papers
RF	Rockefeller Foundation
Sage papers	Russell Sage Foundation papers
Ruml papers	Beardsley Ruml papers, University of Chicago Archives
Seligman papers	Edwin R. A. Seligman papers, Rare Book and Manuscript Library, Columbia University, New York, N.Y.
Tugwell papers	Rexford G. Tugwell papers, Franklin D. Roosevelt Library, Hyde Park, N.Y.
van Kleeck papers	Mary van Kleeck papers, Sophia Smith Collection, Smith College, Northampton, Mass.

Works by Thorstein Veblen

ECO	*Essays in Our Changing Order*. Edited by Leon Ardzrooni. New York: Kelley, 1964.
E&PS	*The Engineers and the Price System*. 1919; reprint, New Brunswick, N.J.: Transaction, 1983.
ER&R	*Essays, Reviews and Reports: Previously Uncollected Writings of Thorstein Veblen*. Edited by Joseph Dorfman. Clifton, N.J.: Kelley, 1973.
PSMC	*The Place of Science in Modern Civilization and Other Essays*. New York: Huebsch, 1919.

TBE	*The Theory of Business Enterprise.* 1904; reprint, New York: Kelley, 1964.
TLC	*The Theory of the Leisure Class: An Economic Study of Institutions.* 1899; reprint, New York: Mentor, 1953.

INTRODUCTION

1. Hawthorne quoted in Leo Marx, *The Machine in the Garden: Technology and the Pastoral Ideal in America* (New York: Oxford University Press, 1964), p. 13.

2. I believe that the term *technocratic progressives* was first used by Robert Westbrook in "Tribune of the Technostructure: The Popular Economics of Stuart Chase," *American Quarterly* 32 (1980): 387–408; political rationalism is discussed in Michael Oakeshott, "Rationalism in Politics," *Cambridge Journal* 1 (1947): 81–98, 145–57.

3. Howard Odum to T. J. Wilson, April 24, 1939, Odum papers. On eugenics, see Daniel Kevles, *In the Name of Eugenics* (New York: Knopf, 1985). For a discussion of class position as a motivation for this style of reform, see Robert Church, "Economists as Experts: The Making of an Academic Profession in the United States, 1870–1920," in *The University in Society*, ed. Lawrence Stone (Princeton: Princeton University Press, 1974), 2:598. On the engineering ideology within a larger framework of class and gender concerns, see Martha Banta's work in progress. Beard is quoted on p. 221 of the present text, Odum on p. 268.

4. JoAnne Brown, "The Semantics of Profession: Metaphor and Power in the History of Psychological Testing, 1890–1929" (Ph.D. diss., University of Wisconsin, 1985), pp. 251, 272. On language and politics in this connection, see also JoAnne Brown, "Professional Language: Words That Succeed," *Radical History Review* 34 (1986): 33–51; David O. Edge, "Technological Metaphor," in *Meaning and Control: Essays in Social Aspects of Science and Technology*, ed. David O. Edge and J. N. Wolfe (London: Tavistock, 1973); and Charles Rosenberg, "Science and American Social Thought," in *Science and Society in the United States*, ed. David Van Tassel and Michael Hall (Homewood, Ill.: Dorsey, 1966), p. 136.

5. In *The Origins of American Social Science* (New York: Cambridge University Press, 1991), pp. 247–53, Dorothy Ross offers an extended discussion of the changing meanings of *social control*. A recent study states that "for the purposes of this analysis, the concept of 'social control' refers narrowly to those practices, from imprisonment to mental hospitalization, by which 'troublesome' populations are removed from the everyday life of society." The people who used the phrase before 1940 seldom had such a precise understanding in mind. See William Staples, *Castles of Our Conscience: Social Control and the American State, 1880–1985* (Cambridge: Polity Press, 1990), p. 155.

6. On the linkage of empirical investigation to large-scale technical systems, see

Samuel Hays's introduction to *Building the Organizational Society*, ed. Jerry Israel (New York: Free Press, 1972), p. 3. Daniel Rodgers identifies the "language of social efficiency" as one of three available "languages of discontent" within the complicated reform atmosphere of the pre–World War I era. Rodgers, "In Search of Progressivism," *Reviews in American History* 10 (1982): 123.

7. Godfrey Hodgson, *America in Our Time* (New York: Vintage, 1976), p. 76; Alan Brinkley, "Roots: The Perotist Tradition," *NR*, July 27, 1992, pp. 44–45.

8. On medium and message, see Paul Corcoran, *Political Language and Rhetoric* (St. Lucia: University of Queensland Press, 1979), pp. 137–49.

9. Hans Morganthau, *Scientific Man Versus Power Politics* (Chicago: University of Chicago Press, 1946), p. 221.

CHAPTER 1

1. On this topic, see Dorothy Ross, *The Origins of American Social Science* (New York: Cambridge University Press, 1991).

2. Ibid., pp. 88–94; Ward quoted on p. 92.

3. Lester Frank Ward, *The Psychic Factors of Civilization* (Boston: Ginn, 1893), p. 309. See also Ward, *Dynamic Sociology* (Boston: Ginn, 1883), 2:249.

4. Lester Frank Ward, review of Thorstein Veblen, *The Theory of the Leisure Class*, reprinted in Veblen, *ER&R*, p. 619.

5. Thorstein Veblen, "The Place of Science in Modern Civilization," in Veblen, *PSMC*, pp. 1, 2; Thorstein Veblen, "Economic Theory in the Calculable Future," in Veblen, *ECO*, p. 7; Thorstein Veblen, *TLC*, p. 131; Thorstein Veblen, "The Instinct of Workmanship and the Irksomeness of Labor," in Veblen, *ECO*, p. 79.

6. Veblen, "Instinct of Workmanship," p. 96; Veblen, *TLC*, pp. 89, 133; T. W. Adorno, "Veblen's Attack on Culture," *Studies in Philosophy and Social Science* 9 (1941): 391.

7. Veblen, *TLC*, p. 75; Veblen, "Instinct of Workmanship," p. 81.

8. Thorstein Veblen, *TBE*, pp. 358, 5, 306; Veblen, *TLC*, pp. 154, 135, 161.

9. Veblen, "Place of Science in Modern Civilization," pp. 17, 3; Veblen, "Economic Theory in the Calculable Future," p. 7; Thorstein Veblen, "Why Is Economics Not an Evolutionary Science?," in Veblen, *PSMC*, p. 67.

10. Veblen, *TLC*, p. 142; Veblen, *TBE*, pp. 372, 311, 309; Veblen, "Place of Science in Modern Civilization," p. 2.

11. Joseph Dorfman, *The Economic Mind in Western Civilization* (New York: Viking, 1949), 3:445; Veblen, *TBE*, pp. 288, 295, 285, 398, 350, 337, 355; Veblen, "Place of Science in Modern Civilization," p. 21; Joseph Dorfman, *Thorstein Veblen and His America* (New York: Viking, 1934), p. 500; Thorstein Veblen, "The Socialist Economics of Karl Marx and His Followers," in Veblen, *PSMC*, pp. 409–56, esp.

pp. 436–37; Thorstein Veblen, review of Thomas Kirkup, *A History of Socialism*, in Veblen, *ER&R*, pp. 419, 420.

12. Dorfman, *Thorstein Veblen*, p. 68; Veblen, *TBE*, pp. 351, 352; Veblen, review of Kirkup, *History of Socialism*, p. 419; Edward Bellamy, *Looking Backward* (1888; reprint, New York: Signet, 1960), p. 54. See also Daniel Aaron, *Men of Good Hope: A Story of American Progressives* (New York: Oxford University Press, 1951); William M. Dugger, "The Origins of Thorstein Veblen's Thought," *Social Science Quarterly* 60 (1979): 422–31; and Joseph Dorfman, "The Source and Impact of Veblen's Thought," in *Thorstein Veblen*, ed. Douglas Dowd (Ithaca: Cornell University Press, 1958), pp. 1–12.

13. Thorstein Veblen, "Some Neglected Points in the Theory of Socialism," in Veblen, *PSMC*, pp. 401, 407; Bellamy, *Looking Backward*, pp. 128, 84; Thorstein Veblen, "The Passing of National Frontiers," in Veblen, *ECO*, pp. 383–90; Thorstein Veblen, *E&PS*, p. 112; Bellamy, *Looking Backward*, pp. 103, 157. On Bellamy and Veblen, see John Thomas, *Alternative America: Henry George, Edward Bellamy, Henry Demarest Lloyd and the Adversary Tradition* (Cambridge, Mass.: Harvard University Press, 1983), pp. 259–60; Aaron, *Men of Good Hope*, p. 217; and, less convincingly, David Riesman, *Thorstein Veblen: A Critical Interpretation* (New York: Scribner's, 1953), p. 9.

14. Thorstein Veblen, "Arts and Crafts," in Veblen, *ECO*, p. 195; Veblen, *TBE*, p. 351; Veblen, "Instinct of Workmanship," p. 85. See also Daniel Horowitz, "Consumption and Its Discontents: Simon N. Patten, Thorstein Veblen, and George Gunton," *Journal of American History* 67 (1980): 301–17.

15. Adolf Berle and Gardiner Means, *The Modern Corporation and Private Property* (New York: Macmillan, 1932), p. 4; Dorfman, "Source and Impact of Veblen's Thought," p. 8; Malcolm Cowley and Bernard Smith, eds., *Books That Changed Our Minds* (1938; reprint, Freeport, N.Y.: Books for Libraries Press, 1970), pp. 3–26, 91–110; Veblen, *ER&R*, pp. 200, 87, 165; Dorfman, *Thorstein Veblen*, pp. 423, 451.

16. On the engineer as hero, see Cecilia Tichi, *Shifting Gears: Technology, Literature, Culture in Modernist America* (Chapel Hill: University of North Carolina Press, 1987), pp. 75–170.

17. Richard Maclurin, *The Mechanic Arts* (Boston: Hall and Locke, 1911), 1:xxv, quoted in Tichi, *Shifting Gears*, p. 97; Tichi, *Shifting Gears*, pp. 98–105; JoAnne Brown, "Professional Language: Words That Succeed," *Radical History Review* 34 (1986): 35–51; H. L. Mencken, *The American Language*, 4th ed. (New York: Knopf, 1945), p. 290. On engineering in this period, see Monte Calvert, *The Mechanical Engineer in America, 1830–1910: Professional Cultures in Conflict* (Baltimore: Johns Hopkins University Press, 1967); A. Michel McMahon, *The Making of a Profession: A Century of Electrical Engineering in America* (New York: IEEE Press, 1984); Bruce Sinclair, *A Centennial History of the American Society of Mechanical*

Engineers, 1880–1980 (Toronto: University of Toronto Press, 1980); and Edwin Layton, *The Revolt of the Engineers: Social Responsibility and the American Engineering Profession* (Cleveland: Case Western University Press, 1971).

18. Tichi, *Shifting Gears,* pp. 100–117; Peter Meiksins, "Professionalism and Conflict: The Case of the American Association of Engineers," *Journal of Social History* 19 (1986): 410; Peter Meiksins, "Scientific Management and Class Relations: A Dissenting View," *Theory and Society* 13 (1984): 200.

19. F. H. Newall and C. E. Drayer, *Engineering as a Career* (New York: Van Nostrand, 1916), pp. 3, 4, 5.

20. George W. Melville, "The Engineer's Duty as a Citizen," *Transactions of the ASME* 32 (1910): 528, 529, 532.

21. Talcott Williams, "Discussion," *Transactions of the ASME* 30 (1908): 630; Melville, "Engineer's Duty," 532.

22. Jean Christie, *Morris Llewellyn Cooke: Progressive Engineer* (New York: Garland, 1983), p. 25. Many city managers were engineers who understood street, trolley, and bridge construction—capital improvements where grafters made so much money. See Barry Karl, *Executive Reorganization and Reform in the New Deal: The Genesis of Administrative Management, 1900–1939* (Cambridge, Mass.: Harvard University Press, 1963), p. 20.

23. Frederick W. Taylor quoted in Lyndall Urwick, *The Golden Book of Management* (London: Management Publications Trust, 1949), p. 60; Frederick W. Taylor, *The Principles of Scientific Management* (1911; reprint, New York: Norton, 1967), p. 5; Samuel Hays, *Conservation and the Gospel of Efficiency* (Cambridge, Mass.: Harvard University Press, 1957), pp. 271, 123, 3.

24. Hays, *Conservation,* pp. 267, 268.

25. Thomas Haskell, *The Emergence of Professional Social Science* (Urbana: University of Illinois Press, 1978), p. 234; Dorothy Ross, "The Social Sciences," in *The Organization of Knowledge in Modern America, 1860–1920,* ed. Alexandra Oleson and John Voss (Baltimore: Johns Hopkins University Press, 1979), p. 122; David A. Hollinger, "Inquiry and Uplift: Late Nineteenth Century American Academics and the Moral Efficacy of Scientific Practice," in *The Authority of Experts: Studies in History and Theory,* ed. Thomas Haskell (Bloomington: Indiana University Press, 1984), p. 143.

26. Albert Shaw, "Presidential Address to the Third Annual Meeting," *APSR* 1 (1907): 184; Raymond Seidelman and Edward Harpham, *Disenchanted Realists: Political Science and the American Crisis, 1880–1980* (Albany: State University of New York Press, 1985), p. 61. On the Wisconsin Plan, see John Rogers Commons, *Myself* (New York: Macmillan, 1934).

27. Robert L. Church, "Economists as Experts: The Making of an Academic Profession in the United States, 1870–1920," in *The University in Society,* ed. Lawrence Stone (Princeton: Princeton University Press, 1974), 2:558; Mary Furner, *Advocacy and Objectivity: A Crisis in the Professionalization of American Social*

Science, 1865–1905 (Lexington: University Press of Kentucky, 1975), pp. 290–91.

28. Dwight Waldo, "Political Science: Tradition, Discipline, Profession, Science, Enterprise," in *The Handbook of Political Science*, ed. Fred Greenstein and Nelson Polsby (Reading, Mass.: Addison-Wesley, 1975), p. 34; Furner, *Advocacy and Objectivity*, pp. 280, 278; Albert Somit and Joseph Tanenhaus, *The Development of American Political Science* (Boston: Allyn and Bacon, 1967), p. 45.

29. Somit and Tanenhaus, *Development of American Political Science*, pp. 66, 83; A. Lawrence Lowell, *Essays on Government* (Boston: Houghton Mifflin, 1889), p. 1, quoted in Bernard Crick, *The American Science of Politics* (London: Routledge and Kegan Paul, 1959), p. 103; A. Lawrence Lowell, "The Physiology of Politics," *APSR* 4 (1910): 4.

30. Woodrow Wilson, "The Law and the Facts," *APSR* 5 (1911): 9, 10; Woodrow Wilson, "Democracy and Efficiency," *Atlantic Monthly* 87 (1901): 297. Another giant of the field concurred. Lord Bryce "denied that political science could ever approximate mechanics or, for that matter, even meteorology." Bryce quoted in Somit and Tanenhaus, *Development of American Political Science*, p. 78.

31. Crick, *American Science of Politics*, p. 105; Arthur Bentley, *The Process of Government* (Chicago: University of Chicago Press, 1908), p. 163. See also Waldo, "Political Science," esp. pp. 37–41.

32. Henry Jones Ford, "The Pretensions of Sociology," *AJS* 9 (1909): 102; anonymous respondent quoted in Furner, *Advocacy and Objectivity*, p. 296; Arthur Vidich and Stanford Lyman, *American Sociology: Worldly Rejections of Religion and Their Directions* (New Haven: Yale University Press, 1985), p. 1; Albion Small, "The Meaning of Sociology," *AJS* 8 (1908): 3 (emphasis in original); Albion Small, "A Vision of Social Efficiency," *AJS* 14 (1914): 433. See also Robert C. Bannister, *Sociology and Scientism: The American Quest for Objectivity, 1880–1940* (Chapel Hill: University of North Carolina Press, 1987).

33. Small, "Meaning of Sociology," p. 14.

34. Ford, "Pretensions," pp. 98, 99, 103.

35. Furner, *Advocacy and Objectivity*, p. 307; Fred Matthews, *Quest for an American Sociology: Robert Park and the Chicago School* (Montreal: McGill-Queen's University Press, 1977), p. 95.

36. Dorothy Ross, "American Social Science and the Idea of Progress," in Haskell, *Authority of Experts*, p. 157.

CHAPTER 2

1. *The Memoirs of Herbert Hoover*, 3 vols. (New York: Macmillan, 1951–52), 1:133; Herbert Hoover to Mary Austin, quoted in George Nash, *The Life of Herbert Hoover: The Engineer, 1874–1914* (New York: Morrow, 1983), p. 482.

2. ASME president E. D. Maier quoted in Edwin Layton, *Revolt of the Engineers*

(Cleveland: Case Western University Press, 1971), p. 60; Herbert Hoover, *Principles of Mining* (New York: McGraw-Hill, 1909), p. 161.

3. Monte Calvert, *The Mechanical Engineer in America, 1830–1910* (Baltimore: Johns Hopkins University Press, 1967).

4. Hoover, *Principles of Mining*, pp. 192–93.

5. Nash, *Herbert Hoover*, p. 490; Herbert Hoover and Lou Henry Hoover, trans., *De Re Metallica* (London: Privately printed, 1912), p. iii.

6. Layton, *Revolt*, pp. 57–67; Bruce Sinclair, *A Centennial History of the American Society of Mechanical Engineers, 1880–1980* (Toronto: University of Toronto Press, 1980), p. 100.

7. On Steinmetz and the National Association of Corporation Schools, see James Gilbert, *Designing the Industrial State* (Chicago: Quadrangle, 1972), pp. 195–99.

8. "Harrington Emerson," *National Cyclopaedia of American Biography* (New York: James T. White, 1916), p. 82; Harrington Emerson, *Efficiency as a Basis for Operation and Wages* (New York: Engineering Magazine, 1909), pp. 171, 15.

9. Richard C. Maclurin quoted in Kenneth Trombley, *The Life and Times of a Happy Liberal* (New York: Harper and Row, 1954), p. 11; Morris L. Cooke, "The Spirit and Significance of Scientific Management," *Journal of Political Economy* 21 (1913): 483, 484.

10. Morris L. Cooke, "Some Factors in Municipal Engineering," *Transactions of the ASME* 36 (1914): 608.

11. Charles Whiting Baker, "Discussion," ibid., p. 628; Robert Wolf, "Discussion," ibid., p. 626. Useful on this point, and on many others, is James T. Kloppenberg, *Uncertain Victory: Social Democracy and Progressivism in European and American Thought, 1870–1920* (New York: Oxford University Press, 1986). Also commenting on the paper was Charles Merriam, who had not yet retired from electoral politics. Charles Merriam, "Discussion," *Transactions of the ASME* 36 (1914): 619.

12. Wolf, "Discussion," p. 627.

13. Morris L. Cooke, "Foreword," *Annals of the American Academy of Political and Social Science* 85 (1919): xi; Morris L. Cooke, "Public Engineering and Human Progress," paper presented before Cleveland Engineering Society, November 14, 1916, pp. 4, 5; Jean Christie, *Morris Llewellyn Cooke—Progressive Engineer* (New York: Garland, 1983), p. 17. Cooke was untiring in his evangelistic efforts. See "The Public Interest as the Bed Rock of Professional Practice," *Transactions of the ASME* 49 (1918): 85–100; "Scientific Management of the Public Business," *APSR* 9 (1915): 488–95; "The Engineer as Citizen," *Journal of the ASME* 41 (1919): 448–51, 496; and *Our Cities Awake* (Garden City, N.Y.: Doubleday, Page, 1919).

14. Lyndall Urwick, *The Golden Book of Management* (London: Management Publications Trust, 1949), p. 31; Taylor quoted on p. 35.

15. The establishment of business schools in the early twentieth century reflected the growing complexity of managerial tasks and a "professionalization" that, like

that of the engineers, did not grant managers the autonomy characteristic of law or medicine. Wharton, founded in 1881, was the first of the lot, followed by, for example, Chicago in 1898; Wisconsin, Tuck, and New York University in 1900; and Harvard, Northwestern, and Pittsburgh in 1908. Raymond A. Kent, *Higher Education in America* (Boston: Ginn, 1930), p. 79.

16. The character of Schmidt, vividly drawn in the book as the "ox-like" worker who outperformed the skeptics after being told how to haul pig iron scientifically, was a fabrication by Taylor based loosely on a worker named Henry Noll, a less than bovine five feet, seven inches and about 135 pounds. The discussion between the "high-priced man" and the nameless manager, so colorfully derogatory as to bring public pressure on Taylor even in 1912, never took place. One can only speculate that Taylor became so obsessed with public approval that he resorted to intense dramatization. His theory was too perfect for the people in it. See Charles Wrege and Armedeo Perroni, "Taylor's Pig Tale: A Historical Analysis of Frederick W. Taylor's Pig Iron Experiments," *Academy of Management Journal* 17 (1974): 6–27.

17. Frank B. Copley, *Frederick Winslow Taylor: Father of Scientific Management* (New York: Harper, 1923), pp. 297, 289.

18. Philippa Strum, *Brandeis* (Cambridge, Mass.: Harvard University Press, 1985), p. ix; Daniel Nelson, *Frederick W. Taylor and the Rise of Scientific Management* (Madison: University of Wisconsin Press, 1980), p. 135; Louis Brandeis to Horace Drury, January 31, 1914, in *The Letters of Louis Brandeis*, ed. David Levy and Melvin Urofsky (Albany: State University of New York Press, 1971), 2:241. For more on Brandeis and Taylor, see Samuel Haber, *Efficiency and Uplift: Scientific Management in the Progressive Era, 1880–1920* (Chicago: University of Chicago Press, 1964), pp. 75–82. This account owes much to Haber's discussion of the entire efficiency crusade.

19. Milton Nadworthy, *Scientific Management and the Unions* (Cambridge, Mass.: Harvard University Press, 1955), p. 36; *New York Times*, November 20, 1910, p. 6, and November 30, 1910, p. 1; *New York Evening Post*, November 26, 1910, quoted in Alpheus T. Mason, *Brandeis: A Free Man's Life* (New York: Viking, 1946), p. 327. The portion of the Brandeis brief devoted to Taylorism was reprinted as *Scientific Management and Railroads* (New York: Engineering Magazine, 1911). See also Oscar Kraines, "Brandeis' Philosophy of Scientific Management," *Western Political Quarterly* 13 (1960): 194.

20. For Brandeis's response to labor opposition, see his introduction to *Primer of Scientific Management*, by Frank Gilbreth (New York: 1912), pp. iii–iv.

21. Luther Gulick, *The Efficient Life* (Garden City, N.Y.: Doubleday, Page, 1911), pp. 93, 35, v.

22. Walter Lippmann, "More Brains, Less Sweat," *Everybody's* 25 (1911): 827–28.

23. Edward Earle Purinton, "Efficiency and Life," *Independent*, November 30, 1914, pp. 321–23.

24. "What Is Efficiency?," *Independent*, November 30, 1914, pp. 326–36. See also Haber, *Efficiency and Uplift*, pp. 51–74.

25. Henry Beech Needham, interview with Theodore Roosevelt, *System* 19 (1913): 586.

26. Barth quoted in Haber, *Efficiency and Uplift*, p. 36.

27. "The Founding of the American Society for Promoting Efficiency," *Transactions of the Efficiency Society* 1 (1912): 22–28; Charles Buxton Going, "The Efficiency Movement: An Outline," ibid., p. 11; Herbert Valentine, "Social Efficiency," ibid., p. 407. The Efficiency Society, in contrast to the Taylorites, included many avowed eugenicists.

28. Henry L. Gantt, "Measuring Efficiency," *Transactions of the ASME* 36 (1914): 417–23. Gantt charts are currently available on computer software.

29. James Hartness, "The Human Element as the Key to Economic Problems," ibid., pp. 384–85.

30. Henry Bruère, "The Future of the Police Arm from an Engineering Standpoint," ibid., pp. 535, 536.

31. *Frederick W. Taylor: A Memorial Volume* (New York: Taylor Society, 1920). See also Layton, *Revolt*.

32. Gantt, *Organizing for Work* (New York: Harcourt, Brace and Howe, 1919), p. 109; Cooke, "Foreword," p. xi.

33. Taylor quoted in Urwick, *Golden Book of Management*, p. 76; Haber, *Efficiency and Uplift*, p. 17.

34. Charles P. Steinmetz, "The Bolsheviks Won't Get You—But You've Got to Watch Out!," *American Magazine*, April 1919, p. 9.

35. Mary B. Mullett, introduction to Steinmetz, "The Bolsheviks," p. 9; Jessica Smith, "Some Memoirs of Russia in Lenin's Time," in *Lenin's Impact on the United States*, ed. Jessica Smith and Daniel Mason (New York: International Publishers, 1970), p. 104.

36. Charles P. Steinmetz, *America and the New Epoch* (New York: Harper, 1916), p. 27 (emphasis in original).

37. Steinmetz, *New Epoch*, pp. 40, 54, 58; New York *Call*, August 21, 1923, pp. 1–2. Veblen had written that "invention is the mother of necessity" in *The Instinct of Workmanship* (1914; reprint, New York: Kelley, 1964), p. 314.

38. Steinmetz, *New Epoch*, pp. 154, 156, 160, 175, 173. See also James Gilbert, "Collectivism and Charles Steinmetz," *Business History Review* 47 (1974): 534.

39. Steinmetz, *New Epoch*, p. 181.

40. Ibid., pp. 62, v.

41. Charles P. Steinmetz, "Presidential Address," *Proceedings of the National Association of Corporation Schools* 3 (1915): 841 (emphasis added). On Steinmetz, see also John M. Jordan, "'Society Improved the Way You Can Improve a Dynamo': Charles P. Steinmetz and the Politics of Efficiency," *Technology and Culture* 30 (1989): 57–82. The standard biography, which appeared as this book was in press,

is Ronald Kline, *Steinmetz: Engineer and Socialist* (Baltimore: Johns Hopkins University Press, 1992).

42. Henry L. Gantt, "Efficiency and Democracy," *Transactions of the ASME* 40 (1918): 799–808. On management, see Henry L. Gantt, *Work, Wages, and Profits*, 2d ed. (New York: Engineering Magazine, 1913). On Gantt's influence on technocracy, see William Akin, *Technocracy and the American Dream* (Berkeley: University of California Press, 1977), pp. 51, 53.

43. Gantt, *Organizing for Work*, pp. iii, 24, 18, 20, 61; Gantt quoted in Charles W. Wood, *The Great Change* (New York: Boni and Liveright, 1918), pp. 43, 44. This book provides a useful compendium of technocratic Progressive thought, with pieces on Veblen, Gantt, and the social worker Mary van Kleeck, among others. See also Gantt, "Efficiency and Democracy," p. 800.

44. "Interview: The Engineer's Formula for Reconciling Capital and Labor on a Basis of Equity," New York Sunday *World*, October 12, 1919, p. 1. See *Industrial Management*, April 1917, p. 127. Walter Polakov included Gantt with Bacon and Descartes in "Gantt: Industrial Leader," *Industrial Management*, December 1919, p. 490. See also Polakov, "Discussion" (of H. L. Gantt, "Efficiency and Democracy"), *Transactions of the ASME* 40 (1918): 807.

45. L. P. Alford, *Henry Laurence Gantt: Leader in Industry* (New York: Harper, 1934), p. 264.

46. Charles Ferguson, "The Men of 1916," *Forum*, February 1916, p. 164. See also Haber, *Efficiency and Uplift*, p. 46.

47. Ferguson, "Men of 1916," pp. 174, 176, 178, 171. See also Charles Ferguson, *The Great News* (New York: Mitchell Kennerly, 1915), pp. 59–64.

48. Frederic Paxson, *American Democracy and the World War* (1936; reprint, New York: Cooper Square, 1966), 1:290. See also David M. Kennedy, *Over Here: The First World War and American Society* (New York: Oxford University Press, 1980).

49. Robert Cuff, *The War Industries Board: Business-Government Relations during World War I* (Baltimore: Johns Hopkins University Press, 1973); C. E. Knoeppel, *Industrial Preparedness* (New York: Engineering Magazine, 1916), pp. 15, 106. See also C. E. Knoeppel, "Industrial Lessons from the German War Machine," *Engineering Magazine*, May 1916, pp. 167–76.

50. C. E. Knoeppel, "The Importance of the Human Factor in Industrial Preparedness," in *The Human Factor in Industrial Preparedness*, by the Western Efficiency Society (Chicago: Holmquist, 1917), p. 37; Cuff, *War Industries Board*, p. 26.

51. O. T. (Ordway Tead), "The Efficiency Movement," *NR*, February 9, 1918, p. 63; Robert F. Hoxie, *Scientific Management and Labor* (New York: Appleton, 1915).

52. Walter Polakov quoted in Wood, *Great Change*, p. 111.

53. Henry P. Kendall, "Discussion," *Bulletin of the Taylor Society* 4 (1919): 18; Henry Shelton and Morris Cooke, "Discussion," ibid., p. 28–29.

54. Kloppenberg, *Uncertain Victory*, p. 363; Grant McConnell, *Private Power and American Democracy* (New York: Knopf, 1966), p. 34.

55. Going, "Efficiency Movement," p. 13.

56. Harry Braverman, *Labor and Monopoly Capital* (New York: Monthly Review Press, 1974).

CHAPTER 3

1. David Croly to Herbert Croly, March 10, 1887, quoted in David Levy, *Herbert Croly of the New Republic* (Princeton: Princeton University Press, 1985), p. 58. See also James T. Kloppenberg, *Uncertain Victory: Social Democracy and Progressivism in European and American Thought, 1870–1920* (New York: Oxford University Press, 1986), p. 315.

2. Herbert Croly, *The Promise of American Life* (New York: Macmillan, 1909), pp. 152, 270, 282, 434, 167; the quotations are from pp. 152, 142, 380, 158. For more on the book, see Levy, *Herbert Croly*, pp. 96–135.

3. Herbert Croly, *Progressive Democracy* (New York: Macmillan, 1914), pp. 25, 426, 376, 399–400, 378–79. See also Levy, *Herbert Croly*, pp. 162–84, and Kloppenberg, *Uncertain Victory*, pp. 383–85.

4. On the origins of the *New Republic*, see Levy, *Herbert Croly*, pp. 185–217.

5. Charles Forcey, *Crossroads of Liberalism* (New York: Oxford University Press, 1961), p. 8; Lippmann quoted in Levy, *Herbert Croly*, p. 190; Johnson quoted in Levy, *Herbert Croly*, p. 192.

6. Walter Weyl, *The New Democracy* (New York: Macmillan, 1912), pp. 315, 313; Levy, *Herbert Croly*, p. 267.

7. John Reed, *The Day in Bohemia* (New York: Privately printed, 1913), p. 42; Walter Lippmann, *A Preface to Politics* (1913; reprint, Ann Arbor: University of Michigan Press, 1962), p. 79.

8. Lippmann, *Preface to Politics*, pp. 22, 182–83, 15, 184. See also John Morton Blum, "Walter Lippmann and the Problem of Order," in *Public Philosopher: Selected Letters of Walter Lippmann*, ed. John Morton Blum (New York: Ticknor and Fields, 1985).

9. Lippmann, *Preface to Politics*, pp. 183, 56, 225, 151.

10. For two extended discussions of *Drift and Mastery*, see William Leuchtenburg's introduction to the Wisconsin edition (see n. 11 below) and David A. Hollinger, "Science and Anarchy: Walter Lippmann's *Drift and Mastery*," in *In the American Province*, by David A. Hollinger (Bloomington: Indiana University Press, 1985), pp. 44–55.

11. Walter Lippmann, *Drift and Mastery* (1914; reprint, Madison: University of Wisconsin Press, 1985), pp. 107, 38, 44.

12. Ibid., pp. 144–45.

13. Ibid., pp. 177, 87, 151.

14. "What Is Opinion?," *NR*, September 18, 1915, p. 171; unsigned, undated memorandum in Lippmann papers.

15. "The Rise of a New Profession," *NR*, May 25, 1918, p. 103; "Democratic Control of Scientific Management," *NR*, December 23, 1916, p. 204.

16. *NR*, December 19, 1914, p. 5; *NR*, May 13, 1916, p. 28; "The Expert and American Society," *NR*, May 4, 1918, p. 67; *NR*, December 23, 1916, p. 205.

17. "The Process of Administration," *NR*, July 6, 1918, p. 281; Charles Beard, "Political Science in the Crucible," *NR*, November 17, 1917, p. 4; John Dewey, "Political Science as a Recluse," *NR*, April 27, 1918, p. 384.

18. Herbert Croly, "The Effect on American Institutions of a Powerful Military and Naval Establishment," *Annals of the American Academy of Political and Social Science* 66 (1916): 167, 171, 164.

19. *NR*, June 8, 1918, p. 160; "For a Living Executive," *NR*, April 13, 1918, pp. 313, 314.

20. William Hard, "Socialistic Coal," *NR*, November 16, 1918, p. 64; "The World in Revolution," *NR*, May 5, 1917, p. 5; "Preliminary to Reconstruction," *NR*, October 12, 1918, p. 305.

21. John Dewey, "What Are We Fighting For?," in *JD:MW*, 11:99.

22. John Dewey, "Conscience and Compulsion," *NR*, July 14, 1917, p. 297 (emphasis added); Kloppenberg, *Uncertain Victory*, pp. 359–85.

23. *NR*, July 28, 1917, p. 344; *NR*, May 26, 1917, p. 92; Herbert Hoover quoted in advertisement, *NR*, October 27, 1917, p. xiii; Edward Eyre Hunt, "Herbert Clark Hoover," *NR*, September 30, 1916, p. 213.

24. C. L. Vestal to the editor, *NR*, February 9, 1918, p. 57.

25. "Efficiency without Inhumanity," *NR*, February 9, 1918, p. 39.

26. C. L. Vestal to the editor, *NR*, February 23, 1918, p. 110. For veiled references to the letters, see "The Process of Administration," *NR*, July 6, 1918, p. 281, and William Hard, "Efficiency and the He-Man," *NR*, March 9, 1918, p. 165.

27. Randolph Bourne, "New Ideals in Business," *Dial*, February 22, 1917, p. 134.

28. For another view of Dewey, see Sidney Kaplan, "Social Engineers as Saviors: Effects of World War I on Some American Liberals," *Journal of the History of Ideas* 17 (1956): 360. Searching for intellectual idols to replace Dewey, Bourne wrote Van Wyck Brooks that in America "I can think of no intellectual effort outside of Veblen's that has not been propaganda of one sort or another." See Eric Sandeen, ed., *The Letters of Randolph Bourne* (Troy, N.Y.: Whitston, 1980), p. 413.

29. Randolph Bourne, "Magic and Scorn," *NR*, December 2, 1916, p. 130; Randolph Bourne, "The State," in *War and the Intellectuals*, ed. Carl Resek (New York: Harper and Row, 1964), p. 71.

30. Randolph Bourne, "Twilight of Idols," in Resek, *War and the Intellectuals*, pp. 59–60.

31. Ibid., p. 60.

32. Ibid., p. 61.

33. Ibid., pp. 62–64.

34. On the BMR, see Jane Dahlberg, *The New York Bureau of Municipal Research* (New York: New York University Press, 1966).

35. See also Donald T. Critchlow, *The Brookings Institution, 1916–1952: Expertise and the Public Interest in a Democratic Society* (DeKalb: Northern Illinois University Press, 1985).

36. Edward House, *Philip Dru: Administrator* (1912; reprint, New York: Huebsch, 1920), pp. 6, 57, 106, 153.

37. Raymond Moley, *The State Movement for Efficiency and Economy* (New York: Columbia University, 1918), p. iv.

38. Frederick Cleveland, "Can Democracy Be Efficient?: The Mechanics of Administration," in *American Problems of Reconstruction*, ed. Elisha Friedman (New York: Dutton, 1918), p. 453.

39. Albert Wright, "Scientific Criteria for Efficient Democratic Institutions," *Scientific Monthly* 6 (1918): 241; P. G. Nutting, "The Principles and Problems of Government," ibid., 8 (1919): 214.

40. Mary van Kleeck on Kellogg, notes to discussion of Taylor Society Dinner, April 28, 1927, Cooke papers. See also John F. McClymer, *War and Welfare: Social Engineering in America, 1890–1925* (Westport, Conn.: Greenwood, 1980); Michael Gordon, "The Social Survey Movement and Sociology in the United States," *Social Problems* 21 (1973–74): 284–98; and Martin Bulmer et al., eds., *The Social Survey in Historical Perspective, 1880–1990* (New York: Cambridge University Press, 1991).

41. Robert Bruère, "Business and Philanthropy," *Harper's*, June 1916, p. 87; Henry Jessup quoted in McClymer, *War and Welfare*, p. 105. See also Roy Lubove, *The Professional Altruist: The Emergence of Social Work as a Career, 1880–1930* (New York: Atheneum, 1969), esp. pp. 157–219.

42. John D. Rockefeller to F. T. Gates, July 27, 1912, Office of the Messrs. Rockefeller files, RG 2f, B 18, F 143, RAC. See also John M. Jordan, ed., "'To Educate Public Opinion': John D. Rockefeller, Jr., and the Origins of Social Scientific Fact-Finding," *New England Quarterly* 64 (1991): 292–97, and David Grossman, "American Foundations and the Support of Economic Research," *Minerva* 20 (1982): 59–83.

43. Verbatim notes of Princeton Conference [of RF trustees], pp. 163, 164, RG 3, S 900, B 22, F 167, RF papers, RAC. See also the founding document, "An Institute for Government Research," RG 1.1, S 200, B 26, F 295, RF papers, RAC. For a discussion of the relations between private and public sector reform in this period, see Peter Dobkin Hall, "A Historical Overview of the Private Nonprofit Sector," in *The Nonprofit Sector: A Research Handbook*, ed. Walter Powell (New Haven: Yale University Press, 1987), pp. 3–26, esp. 10–15.

44. Lucy Sprague Mitchell, *Two Lives: The Story of Wesley Clair Mitchell and Myself* (New York: Simon and Schuster, 1953), p. 290; Wesley C. Mitchell, "Statistics and Government," *Journal of the American Statistical Association* 125 (March 1919): 229. Mitchell told Lucy Sprague that he wanted to marry her not only "because it would turn the desires which trouble me now into sources of ecstasy, but because it would bring days of serene happiness and a sense of heightened efficiency." Mitchell to Sprague, November 26, 1911, in Lucy Sprague Mitchell papers. He wrote Graham Wallas in 1915 that "I should hail it as a fine result if England suddenly made up her mind to replenish her wasted population from the best breeding stock that she has left, quite without reference to traditional ideas concerning the family." Mitchell to Wallas, February 3, 1915, Wesley C. Mitchell papers.

45. F. Stuart Chapin, "What Is Sociology," *Scientific Monthly* 7 (1918): 262; Charles Ellwood, "Making the World Safe for Democracy," ibid., pp. 514–15, 524; L. L. Bernard, "The Transition to an Objective Standard of Social Control," *AJS* 16 (1911): 532, 534.

46. Jesse Macy, "The Scientific Spirit in Politics," *APSR* 11 (1917): 2–8.

47. See Dorothy Ross, *The Origins of American Social Science* (New York: Cambridge University Press, 1991), esp. p. 388.

CHAPTER 4

1. On the war, see Sidney Kaplan, "Social Engineers as Saviors: Effects of World War I on Some American Liberals," *Journal of the History of Ideas* 17 (1956): 347–69; David M. Kennedy, *Over Here: The First World War and American Society* (New York: Oxford University Press, 1980); John A. Thompson, *Reformers and War: American Progressive Publicists and the First World War* (New York: Cambridge University Press, 1987); James A. Neuchterlein, "The Dream of Scientific Liberalism: *The New Republic* and American Progressive Thought, 1914–1920," *Review of Politics* 42 (1980): 167–90.

2. Robert Cuff, *The War Industries Board: Business-Government Relations during World War I* (Baltimore: Johns Hopkins University Press, 1973), pp. 28, 41.

3. Grosvenor Clarkson, *Industrial America in the World War* (Boston: Houghton Mifflin, 1923), pp. 21–22.

4. Robert Cuff, "We Band of Brothers—Woodrow Wilson's War Managers," *Canadian Review of American Studies* 2 (1974): 137, 138, 143. See also William Leuchtenburg, "The New Deal and the Analogue of War," in *Change and Continuity in Twentieth-Century America*, ed. John Braeman et al. (New York: Harper and Row, 1964), pp. 81–144.

5. Baruch quoted in Kennedy, *Over Here*, p. 130. On Baruch, see also Jordan Schwartz, *The Speculator* (Chapel Hill: University of North Carolina Press, 1981).

6. "War Activities of Persons Associated with the Late Frederick W. Taylor," *Bulletin of the Taylor Society* 4, supplement no. 3 (1919); Herbert Heaton, *A Scholar in Action: Edwin F. Gay* (1942; reprint, New York: Greenwood, 1968), pp. 97–102.

7. Heaton, *Gay*, pp. 112–23.

8. Ibid., pp. 124–36, 168.

9. Lawrence Gelfand, *The Inquiry* (New Haven: Yale University Press, 1963), pp. 38, 314–15; Kaplan, "Social Engineers," p. 359.

10. Walter Lippmann, *The Stakes of Diplomacy* (New York: Holt, 1915), pp. 199, 201.

11. Walter Lippmann, *Liberty and the News* (New York: Harcourt, Brace, and Howe, 1920), pp. 5, 55, 64, 91–92.

12. Lippmann to Hoover, May 15, 1917, Lippmann papers.

13. See also Robert Cuff, "Herbert Hoover: The Ideology of Voluntarism and War Organization during the Great War," *Journal of American History* 64 (1977): 358–72.

14. David Burner, *Herbert Hoover: A Public Life* (New York: Knopf, 1978), pp. 96–110. See also Herbert Hoover, introduction to *History of the United States Food Administration, 1917–1919*, by William Mullendore (Palo Alto: Stanford University Press, 1941), pp. 3–41.

15. Hoover to Mary Austin, April 11, 1918, HHPL.

16. Terman quoted in Daniel Kevles, *In the Name of Eugenics: Genetics and the Uses of Human Heredity* (1985; reprint, Berkeley: University of California Press, 1986), p. 81; George Patrick, "The Psychology of Social Reconstruction," *Scientific Monthly* 6 (1918): 508.

17. Devine quoted in John F. McClymer, *War and Welfare: Social Engineering in America, 1890–1925* (Westport, Conn.: Greenwood, 1980), p. 172; Charles Ellwood, "The Educational Theory of Social Progress," *Scientific Monthly* 5 (1917): 450. See also David Levine, *The American College and the Culture of Aspiration, 1915–1940* (Ithaca: Cornell University Press, 1986).

18. "The Uses of an Armistice," *NR*, November 6, 1918, p. 60; Charles W. Wood, *The Great Change* (New York: Boni and Liveright, 1918), pp. 31–32; Wilson quoted in Robert Bruère, "Changing America," *Harper's*, February 1919, p. 289; Charles Ferguson, "The Men of 1916," *Forum*, February 1916, p. 148. On the new and the modern, see David A. Hollinger, "The Knower and the Artificer," *American Quarterly* 39 (1987): 37–55.

19. Raymond Fosdick, *The Old Savage in the New Civilization* (Garden City, N.Y.: Doubleday, Doran, 1931), pp. 27–28.

20. Wood, *Great Change*, p. 20; Robert Bruère, "The New Nationalism and Business," *Harper's*, March 1919, p. 511.

21. Clarkson, *Industrial America*, p. 312; Rexford Tugwell, "America's War-Time Socialism," *Nation*, April 6, 1927, pp. 364, 365, 367.

22. Herbert Croly, "A School for Social Research," *NR*, June 8, 1918, pp. 167,

168, 170. See also David Levy, *Herbert Croly of the New Republic* (Princeton: Princeton University Press, 1985), pp. 270–71.

23. Butler quoted in Ellen Nore, *Charles Beard: An Intellectual Biography* (Carbondale: Southern Illinois University Press, 1983), p. 89; Emily James Putnam, "The New School of Social Research: A Communication," *NR*, February 4, 1920, p. 294. See also Peter Rutkoff and William Scott, *New School: A History of the New School for Social Research* (New York: Free Press, 1986).

24. Herbert Croly, introduction to *Social Discovery*, by Eduard Lindeman (New York: Republic, 1925), pp. xvii, xiii; A. C. Freeman, "Why Mr. Hoover?," *NR*, March 3, 1920, p. 33. On Hoover, see the unsigned *NR* pieces: "Hoover as President," January 21, 1920, pp. 207–8; "Hoover and the Issues," February 4, 1920, pp. 281–83; "The Meaning of Hoover's Candidacy," February 21, 1920, p. 328; "Fighting for Hoover," March 3, 1920, pp. 4–6; "Hoover's Chances," April 14, 1920, p. 196; and "Concerning Mr. Hoover," May 12, 1920, p. 326.

25. Herbert Croly, "Regeneration," *NR*, June 9, 1920, p. 47; Herbert Croly, "LaFollette," *NR*, October 29, 1924, p. 224; Herbert Croly, "The *New Republic* Idea," insert to *NR*, December 6, 1922, p. 16.

26. Philip Moore, "The Social Responsibility of the Engineer," *Journal of the ASME* 41 (1919): 448; Franklin K. Lane, "A Direct Message for American Engineers," *Industrial Management*, February 1920, pp. 96a–96b; Samuel Gompers to ASME, November 5, 1920, in Pre-Commerce, FAES file, HHPL.

27. William Henry Smyth, "'Technocracy'—National Industrial Management," *Industrial Management*, March 1919, pp. 208, 211, 212 (all emphases in original).

28. Thorstein Veblen, *E&PS*, pp. 84, 86, 90; Samuel Haber, *Efficiency and Uplift: Scientific Management in the Progressive Era, 1880–1920* (Chicago: University of Chicago Press, 1964), p. 143; Daniel Bell, introduction to Veblen, *E&PS*, p. 35. See also Edwin Layton, "Veblen and the Engineers," *American Quarterly* 14 (1962): 64–72.

29. David Riesman, *Thorstein Veblen: A Critical Interpretation* (New York: Scribner's, 1953), p. 46; Lewis Mumford, review of Veblen, *E&PS*, *The Freeman*, November 23, 1921, p. 261; Veblen, *E&PS*, pp. 44, 47, 133, 151.

30. Daniel Aaron, *Men of Good Hope: A Story of American Progressives* (New York: Oxford University Press, 1951), p. 223; Mumford, review of Veblen, *E&PS*, p. 262; C. Wright Mills, introduction to Veblen, *TLC*, p. xvii; T. W. Adorno, "Veblen's Attack on Culture," *Studies in Philosophy and Social Science* 9 (1941): 401. Loren Baritz has pointed out that the very engineers Veblen idolized held assumptions based on the same classical economics he tried to extirpate from his discipline. See Baritz, *The Servants of Power: A History of the Use of Social Science in American Industry* (Middletown, Conn.: Wesleyan University Press, 1960), p. 30.

31. Clarkson, *Industrial America*, p. 9.

32. Walter Lippmann to Graham Wallas, August 31, 1920, in *Public Philosopher: Selected Letters of Walter Lippmann*, ed. John Morton Blum (New York: Ticknor and

Fields, 1985), p. 137; Walter Lippmann, *Public Opinion* (New York: Harcourt, Brace, 1922), pp. 378, 370, 373, 399. On *Public Opinion*, see Edward Purcell, *The Crisis of Democratic Theory: Scientific Naturalism and the Problem of Value* (Lexington: University Press of Kentucky, 1973), pp. 104–7.

33. Walter Lippmann, *The Phantom Public: A Sequel to "Public Opinion"* (New York: Macmillan, 1925), pp. 103, 198–99.

CHAPTER 5

1. Edwin Layton, *Revolt of the Engineers* (Cleveland: Case Western University Press, 1971), p. 179; David Burner, *Herbert Hoover: A Public Life* (New York: Atheneum, 1984), p. 138; David Levy and Melvin Urofsky, eds., *The Letters of Louis Brandeis* (Albany: State University of New York Press, 1975), 4:448; Robert Wiebe, *The Search for Order* (New York: Hill and Wang, 1967); Anne O'Hare McCormick, "America at Last Airs Its Mind," *New York Times Magazine*, November 4, 1928, p. 2.

2. "Hoover as President," *NR*, January 21, 1920, p. 208.

3. Burner, *Hoover*, pp. 152, 154.

4. Ibid., p. 141; George Nash, *The Life of Herbert Hoover: The Engineer, 1871–1914* (New York: Morrow, 1983), p. 39. For one view of Hoover's social thought in this period, see Gary Dean Best, *The Politics of American Individualism* (Westport, Conn.: Greenwood, 1975); on Hoover's "linguistic synthesis," see David Green, *Shaping Political Consciousness: The Language of Politics in America from McKinley to Reagan* (Ithaca: Cornell University Press, 1987), esp. pp. 95–118.

5. Herbert Hoover, *American Individualism* (Garden City, N.Y.: Doubleday, Page, 1922), p. 19. Hoover's reputation probably inspired some reviewers to overreact; the *New York Times* lumped Hoover with Hamilton, Madison, Jay, and Webster while predicting that *American Individualism* "doubtless will rank among the few great formulations of political theory." *New York Times Book Review*, December 17, 1922, p. 1.

6. Herbert Hoover, *On Growing Up* (New York: William Morrow, 1962), pp. 36, 19; *The Memoirs of Herbert Hoover*, 3 vols. (New York: Macmillan, 1951–52), 1:6, 7.

7. Hoover to American Engineering Council, February 14, 1921, PS 128, HHPL; Hoover, *Memoirs*, 1:131; Hoover to American Institute of Electrical Engineers, January 24, 1925, PS 440, and Hoover, "Some Notes on Industrial Readjustment," December 27, 1919, PS 39, both in HHPL.

8. Robert Cuff, "Herbert Hoover: The Ideology of Voluntarism and War Organization during the Great War," *Journal of American History* 64 (1977): 359; Guy Alchon, *The Invisible Hand of Planning: Capitalism, Social Science, and the State in the 1920s* (Princeton: Princeton University Press, 1985), p. 4; Hoover to Associated Advertising Clubs, May 4, 1925, PS 483, and Hoover to University of North Carolina, August 25, 1921, PS 168, both in HHPL.

9. Hoover to Los Angeles *Daily News*, July 13, 1925, PS 501, HHPL; "The Republican Nominee," *Review of Reviews*, July 1928, p. 30; Morris L. Cooke to Charles Merriam, July 30, 1928, B 27, F 16, Merriam papers. See also Peri Arnold, "The 'Great Engineer' as Administrator: Herbert Hoover and Modern Bureaucracy," *Review of Politics* 42 (1980): 329–48.

10. Hoover, *Memoirs*, 2:279; "The Republican Nominee," p. 30; Barry D. Karl, "Presidential Planning and Social Science Research: Mr. Hoover's Experts," *Perspectives in American History* 3 (1969): 361, 363.

11. Hoover press release, January 2, 1933, cited in Hoover, *Memoirs*, 2:313; Herbert Hoover, *Principles of Mining* (New York: McGraw-Hill, 1909), p. 167; American Arbitration Association 1927 Yearbook, PS 765, HHPL.

12. On Hoover and public relations, see Craig Lloyd, *Aggressive Introvert* (Columbus: Ohio State University Press, 1972).

13. See Ellis W. Hawley, "Herbert Hoover, the Commerce Secretariat and the Vision of an 'Associative State,' 1921–1928," *Journal of American History* 62 (1974): 116–40; and Alchon, *Invisible Hand*, p. 77. E. E. Hunt, "The Cooperative Committee Conference System," December 14, 1926, Commerce papers, HHPL.

14. On Hoover and the FAES, see Layton, *Revolt*, chap. 8.

15. William Barber, *From New Era to New Deal* (New York: Cambridge University Press, 1985), p. 14.

16. Hoover to American Engineering Council, January 10, 1925, PS 436A, and Hoover to National Civic Federation, April 11, 1925, PS 468, both in HHPL.

17. Hoover to American Engineering Council, February 14, 1921, PS 128, HHPL; Hoover, *Memoirs*, 2:312; Hoover to Western Society of Engineers, June 2, 1926, PS 591, HHPL (emphasis added).

18. Wesley C. Mitchell, "Statistics and Government," *Journal of the American Statistical Association* 125 (March 1919): 230; Karl, "Presidential Planning," p. 350.

19. Alchon, *Invisible Hand*, pp. 77, 22, 135.

20. As of 1922, for example, Hoover sat on the board of the IGR; among his fellow members were the former war administrators Robert Brookings, Frederick Delano, Raymond Fosdick, and Edwin Gay. The best source on Hoover's coordination of social science, private philanthropy, and public policy is Alchon, *Invisible Hand*, on which I rely heavily in this chapter.

21. Hoover, *Memoirs*, 2:312; Hoover to American Engineering Council, January 10, 1924, PS 345A, HHPL; Herbert Hoover, foreword to *Recent Social Trends*, by President's Research Committee on Social Trends (New York: McGraw-Hill, 1932).

22. John Dunlap, "A Word about Herbert Hoover," *Industrial Management*, January 1, 1921, p. 1; Hoover Presidential Secretary's File—Engineers' National Hoover Committee, HHPL; Hoover, *Memoirs*, 1:133; Hoover, foreword to *America and the New Era*, by Elisha Friedman (New York: E. P. Dutton, 1921), p. xxiv.

23. Kathleen Norris, "A Woman Looks at Hoover," *Collier's*, May 5, 1928, p. 8.

24. Burner, *Hoover*, p. 44; Sullivan quoted (no citation) in Eugene Lyons, *Herbert*

Hoover (New York: Doubleday, 1964), p. 160; E. W. Rice press release, n.d., Campaign Literature box, Campaign and Transition file, HHPL; Dunlap, "A Word," p. 1; Zay Jeffries to "Fellow Engineer," n.d., B 110, Post-Presidential file, HHPL.

25. Hoover to *Chicago Daily News*, March 16, 1922, public statements appendix, HHPL.

26. Hoover to American Institute of Mining and Metallurgical Engineers, February 17, 1920, PS 45, and Hoover to American Engineering Council, January 10, 1924, PS 345A, both in HHPL.

27. Hoover to American Engineering Council, February 14, 1921, PS 129; Hoover to Associated Advertising Clubs, May 11, 1925, PS 483; Hoover to Stanford University graduation, June 22, 1925, PS 499; and Hoover to General Federation of Women's Clubs, May 18, 1925, PS 481, all in HHPL.

28. Hoover to Penn College graduation, June 12, 1925, PS 496, HHPL; Hoover, *American Individualism*, pp. 3, 48.

29. Hoover to American Institute of Electrical Engineers, January 24, 1925, PS 440, HHPL.

30. Hoover, foreword to Friedman, *America and the New Era*, p. xxiii.

31. William Appleman Williams, "What This Country Needs . . . The Shattered Dream: Herbert Hoover and the Great Depression," *New York Review of Books*, November 5, 1970; Walter Lippmann to Felix Frankfurter, November 8, 1930, in *Public Philosopher: Selected Letters of Walter Lippmann*, ed. John Morton Blum (New York: Ticknor and Fields, 1985), p. 268.

32. Hoover to American Engineering Council, January 5, 1922, PS 196; Hoover to American Institute of Mechanical Engineers, February 16, 1921, PS 129; and Hoover to Howard Heinz, October 9, 1924, Public Statements appendix, all in HHPL.

33. Hoover to Rutgers College commencement, June 14, 1920, PS 70, and Hoover to Stanford University graduation, June 22, 1925, PS 499, both in HHPL.

34. Hoover, "Some Notes on Industrial Readjustment," December 27, 1919, PS 39, HHPL; Hoover, foreword to Friedman, *America and the New Era*, p. xxiv; Hoover to Round Table Conference, National Civic Federation, April 11, 1925, PS 468, HHPL; Carl Degler, "Herbert Hoover," *Yale Review* 52 (1963): 580.

35. Walter Lippmann to Herbert Croly, March 24, 1930, in Blum, *Public Philosopher*, p. 260; Hoover, *Memoirs*, 2:136.

36. Morris Cohen, review of Hoover, *American Individualism*, NR, February 21, 1923, p. 353.

37. Ibid.

38. Hoover to Wesley Mitchell, October 17, 1934, quoted in Lucy Sprague Mitchell, *Two Lives: The Story of Wesley Clair Mitchell and Myself* (New York: Simon and Schuster, 1953), p. 371.

39. Ibid.; Burner, *Hoover*, p. 141.

CHAPTER 6

1. Harry Elmer Barnes, introduction to *History and Prospects of the Social Sciences*, ed. Harry Elmer Barnes (New York: Knopf, 1925), p. xv; Knight Dunlap, "The Social Need for Scientific Sociology," *Scientific Monthly* 11 (1920): 516.

2. Harry Elmer Barnes, "Some Contributions of Sociology to Modern Political Theory," in *A History of Political Theories*, ed. Harry Elmer Barnes and Charles Merriam (New York: Macmillan, 1924), p. 373; E. L. Thorndike, "Intelligence and Its Uses," *Harper's*, January 1920, p. 235; A. B. Wolfe, *Conservatism, Radicalism, and Scientific Method* (New York: Macmillan, 1923), p. 219.

3. On scientism in the 1920s, see Dorothy Ross, *The Origins of American Social Science* (New York: Cambridge University Press, 1991), esp. pp. 390–470. For other thoughtful discussions of this period, see also Guy Alchon, *The Invisible Hand of Planning: Capitalism, Social Science, and the State in the 1920s* (Princeton: Princeton University Press, 1985); Robert C. Bannister, *Sociology and Scientism: The American Quest for Objectivity, 1880–1940* (Chapel Hill: University of North Carolina Press, 1987); Barry D. Karl, *Charles E. Merriam and the Study of Politics* (Chicago: University of Chicago Press, 1974), and "Presidential Planning and Social Science Research: Mr. Hoover's Experts," *Perspectives in American History* 3 (1969): 347–407; Edward Purcell, *The Crisis of Democratic Theory: Scientific Naturalism and the Problem of Value* (Lexington: University Press of Kentucky, 1973), esp. pp. 95–114; and Raymond Seidelman and Edward Harpham, *Disenchanted Realists: Political Science and the American Crisis, 1880–1980* (Albany: State University of New York Press, 1985).

4. James T. Shotwell, *Intelligence and Politics* (New York: Century, 1921), pp. 21, 26, 27, 30; Charles Merriam, "The Present State of the Study of Politics," *APSR* 15 (1921): 183–84; Charles Merriam, *New Aspects of Politics* (1925; reprint, Chicago: University of Chicago Press, 1970), pp. 293, 322. See also Dwight Waldo, "Political Science: Tradition, Discipline, Profession, Science, Enterprise," in *The Handbook of Political Science*, ed. Fred Greenstein and Nelson Polsby (Reading, Mass.: Addison-Wesley, 1975), p. 49.

5. Lundberg quoted in Purcell, *Crisis*, p. 20; Barnes, introduction to *History and Prospects of the Social Sciences*, p. xiii; Franklin H. Giddings, *The Scientific Study of Human Society* (Chapel Hill: University of North Carolina Press, 1924), pp. 165, 166, 167. Giddings made a similar argument in "Social Work and Societal Engineering," *Social Forces* 3 (1924): 13–14. See also Arthur Vidich and Stanford Lyman, *American Sociology: Worldly Rejections of Religion and Their Directions* (New Haven: Yale University Press, 1985), pp. 113, 123–24.

6. Alvin Hansen, "Social Scientist and Social Counselor," in *Wesley Clair Mitchell: The Economic Scientist*, ed. Arthur Burns (New York: NBER, 1952), p. 319; Wesley C. Mitchell, review of Herbert Hoover, *Challenge to Liberty*, *Political Science*

Quarterly 49 (1935): 599; Jacob Hollander, "The Economist's Spiral," *American Economics Review* 12 (1922): 12; Ross, *Origins*, p. 415.

7. A. F. Kuhlman, "The Social Science Research Council," *Social Forces* 6 (1928): 583–84.

8. On philanthropic agenda-setting, see Barry D. Karl and Stanley N. Katz, "The American Private Philanthropic Foundation and the Public Sphere, 1890–1930," *Minerva* 19 (1981): 236–70, esp. 237, 248, 259; and Alchon, *Invisible Hand*.

9. My analysis closely follows that of Guy Alchon, Ellis Hawley, and Barry Karl. Because of the extensiveness and quality of their accounts of this period, I have attempted to re-cover their ground as seldom as possible. Keppel quoted in Purcell, *Crisis*, p. 28.

10. Rockefeller quoted in Raymond Fosdick, *A Philosophy for a Foundation* (New York: Rockefeller Foundation, 1963), p. 5; Ellen Condliffe Lagemann, *The Politics of Knowledge: The Carnegie Corporation, Philanthropy, and Public Policy* (Middletown, Conn.: Wesleyan University Press, 1989), p. 68.

11. On the American Association for Labor Legislation, see Donald J. Murphy, "John B. Andrews, the American Association for Labor Legislation, and Unemployment Reform, 1914–1929," in *Voluntarism, Planning, and the State: The American Planning Experience, 1914–1946*, ed. Jerold Brown and Patrick Reagan (Westport, Conn.: Greenwood, 1988), pp. 1–23.

12. *The Institute for Governmental Research: Its Organization, Work, and Publications* (Washington: Institute for Governmental Research, 1922), p. 3; Harold Moulton, "The Brookings Institution: Its Objectives and Its Spirit," S 3.5, B 49, F 518, LSRM papers, RAC.

13. Raymond Fosdick, *The Old Savage in the New Civilization* (Garden City, N.Y.: Doubleday, Doran, 1931), p. 21. See also Fosdick, *Chronicle of a Generation: An Autobiography* (New York: Harper, 1958), p. 275.

14. Report of LSRM, October 18–December 31, 1922, RG 2, B 53, F 33, LSRM papers, RAC. On the understanding of John D. Rockefeller, Jr., of the fund and its initial donations, see "Statement by John D. Rockefeller, Jr. published on Thanksgiving Day," November 24, 1920, S 2, Sub 3, F 33, LSRM papers, RAC. On Ruml, see Martin Bulmer and Joan Bulmer, "Philanthropy and Social Science in the 1920s: Beardsley Ruml and the Laura Spelman Rockefeller Memorial, 1922–29," *Minerva* 19 (1981): 347–407.

15. Ruml memo of October 1922, p. 10, S 2, B 2, F 31, LSRM papers, RAC.

16. Beardsley Ruml, "Social Science in Retrospect and Prospect," in *Eleven Twenty-Six: A Decade of Social Science Research*, ed. Louis Wirth (Chicago: University of Chicago Press, 1940), p. 23; "Memorial Policy in the Social Sciences," p. 10, RG 3, S 910, B 2, F 10, RF papers, RAC (emphasis added).

17. Ruml to E. R. A. Seligman, May 23, 1927, S 3, Sub 5.5, B 51, F 540, LSRM papers, RAC. The economist Lawrence Frank, who published in the *Dial* and the *New Republic* and served as the business manager of the New School for Social Re-

search, worked for the LSRM from 1923 until 1930. When his articles appeared in professional and lay journals, no Rockefeller identification appeared with his name.

18. SSRC 1928–29 annual report. The "Definition of Council Objectives," dated October 1929, listed "enhancement of the public appreciation of the significance of the social sciences" as one of seven goals for the organization. Copy in SSRC-Lynd file, B 34, Merriam papers.

19. Appelget to Fosdick, December 12, 1927, S 3.5, B 50, F 529, LSRM papers, RAC.

20. Vincent to Fosdick and Pritchett to Fosdick, January 5, 1928, both in S 3.5, B 50, F 530, LSRM papers, RAC.

21. "Principles governing the Memorial's program in the social sciences," November 23, 1928, RG 3, S 910, B 2, F 11, RF papers, RAC.

22. Day to RF social science staff conference, January 14, 1930, p. 1, RG 3, S 910, B 1, F 2, RF papers, RAC.

23. Day quoted in transcripts from Princeton conference with RF trustees, October 29, 1930, pp. 118, 129, 145, 150, 152, RG 3, S 900, B 22, F 167; "New Social Science Program," RG 3, S 910, B 2, F 13, both in RF papers, RAC.

24. Merriam, *New Aspects of Politics*, p. 51; Merriam memo of December 27, 1926, p. 24, S 3.5, B 64, F 687, LSRM papers, RAC.

25. Merriam, "Present State of the Study of Politics," pp. 183–84; Merriam to Cooke, October 21, 1924, Merriam papers; SSRC 1926 annual report, p. 15.

26. "National Conference on the Science of Politics," p. 1, B 30, F 21, Merriam papers.

27. SSRC 1926 annual report, p. 14; Merriam, *New Aspects of Politics*, pp. 293, 322.

28. Wesley C. Mitchell, "Institutes for Research in the Social Sciences," reprinted in *The Backward Art of Spending Money and Other Essays*, by Wesley C. Mitchell (New York: McGraw-Hill, 1937), p. 71; Wesley C. Mitchell, "Economics and Social Engineering," reprinted in *Science and Social Change*, ed. Jesse Thornton (Washington, D.C.: Peabody Institute, 1939), p. 313.

29. Mitchell to *New York Evening Post*, March 31, 1920, manuscript letter, and Mitchell to Norwalk Congregational Church, January 27, 1920, p. 6, both in Wesley C. Mitchell papers; Wesley C. Mitchell, "The Prospects of Economics," reprinted in Mitchell, *Backward Art*, pp. 372, 342.

30. Mitchell, "The Trend of Economic Research in Industry," December 14, 1925, handwritten ms., Wesley C. Mitchell papers.

31. Mitchell to Raymond Fosdick, January 18, 1927, ibid.

32. William F. Ogburn, "The Folk-Ways of a Scientific Sociology," *Scientific Monthly* 30 (1930): 300, 301, 303, 306.

33. Howard Odum, introduction to *American Masters of Social Science*, ed. Howard Odum (1927; reprint, Port Washington, N.Y.: Kennikat Press, 1965), p. 9; Howard Odum, *Sociology and Social Problems* (Chicago: American Library Associ-

ation, 1925), p. 18; Wayne Brazil, *Howard Odum: The Building Years, 1884–1930* (New York: Garland, 1988), p. 565.

34. Hornell Hart, *The Techniques of Social Progress* (New York: Holt, 1931), pp. 673, 666.

CHAPTER 7

1. Ruml to 1925 Hanover conference, p. 4, B 135, F 1, Merriam papers. See also SSRC Programs and Policy committee document, p. 2, S 3.6, B 65, F 694, LSRM papers, RAC.

2. For attendance rosters, see S 3.5, B 52, F 562; B 65, F 694; and B 66, F 698 and 701, LSRM papers, RAC; and B 135, F 1, Merriam papers. Merriam tried without success to lure Herbert Croly and Walter Lippmann into visiting the first conference. Merriam to Guy Stanton Ford, July 20, 1925, S 3.5, B 52, F 562, LSRM papers, RAC.

3. Lynd to Howard Odum, February 20, 1930, Odum papers; Merriam, "Impressions of Hanover," p. 6, B 136, F 4, Merriam papers (neither "Impressions of Hanover" nor the similar reflection "Impressions of Hanover Conference" is dated).

4. A. B. Hall to 1926 Hanover conference, p. 7, S 3.5, B 65, F 694, LSRM papers, RAC; Martin Bulmer and Joan Bulmer, "Philanthropy and Social Science in the 1920s: Beardsley Ruml and the Laura Spelman Rockefeller Memorial, 1922–29," *Minerva* 19 (1981): 381–82.

5. Harold Laski, "Foundations, Universities, and Research," reprinted in *The Dangers of Obedience and Other Essays*, by Harold Laski (New York: Harper's, 1930), pp. 154, 158, 162–63. The original article appeared on pp. 295–303 of the magazine. Laski's own institution was a major beneficiary of Rockefeller largess. His colleague R. H. Tawney, to whom *The Dangers of Obedience* was dedicated, received Rockefeller support for the publication of *Religion and the Rise of Capitalism*. See Bulmer and Bulmer, "Ruml," p. 395.

6. Laski, "Foundations," pp. 177, 175, 176.

7. Harold Moulton to 1928 Hanover conference, pp. 393, 394, S 3.6, B 67, F 703, LSRM papers, RAC.

8. Frankfurter to Raymond Fosdick, November 8, 1928, S 3.6, B 66, F 701, ibid.

9. Mitchell to 1926 Hanover conference, S 3.6, B 66, F 697, and Merriam to 1927 Hanover conference, S 3.6, B 66, F 699, both in ibid.

10. Robert MacIver, "Additional Report #2 to SSRC Committee on Scientific Method," p. 56, B 138, F 11, Merriam papers. The published report was *Methods in Social Science: A Casebook*, ed. Stuart Rice (Chicago: University of Chicago Press, 1931).

11. Mr. (?) Peterson to 1927 Hanover conference, p. 53, S 3.6, B 66, F 699, and

F. H. Allport to 1926 Hanover conference, p. 458, S 3.6, B 66, B 697, both in LSRM papers, RAC.

12. F. H. Allport to 1926 Hanover conference, p. 458, S 3.6, B 66, F 697, ibid.; Merriam, "Impressions of Hanover," p. 6; Day to 1928 Hanover conference, pp. 74, 75, S 3.6, B 67, F 702, LSRM papers, RAC.

13. Merriam, "Impressions of Hanover," p. 9.

14. W. F. Ogburn to 1928 Hanover conference, pp. 71, 72, S 3.6, B 67, F 702, LSRM papers, RAC.

15. Frank Knight, "The Limitations of Scientific Method in Economics," reprinted in *The Ethics of Competition and Other Essays*, by Frank Knight (London: Allen and Unwin, 1935), pp. 105–17.

16. Ibid., pp. 117–33.

17. Ibid., pp. 133, 135, 147.

18. The transcript of Knight's paper and the subsequent discussion is found in pp. 137–226, S 3.6, B 67, F 702, LSRM papers, RAC. The adjournment time is noted on p. 226, while interruptions can be found on pp. 153–58, 161, 166, and 168. Quotations from pp. 137–41.

19. Ibid., pp. 158, 468–73.

20. Ibid., p. 468.

21. Merriam, "Impressions of the Dartmouth Conference of Economists," S 3.6, B 66, F 701, LSRM papers, RAC. The copy of Day's Hanover address, "Trends of Social Science," in the Day papers is undated. Other evidence shows that it was given in 1928.

22. Day, "Trends of Social Science," pp. 3, 5, 10.

23. Ibid., pp. 11, 5.

24. A. B. Hall to confidential Hanover conference on future SSRC policy, August 26, 1929, p. 12. (Arthur M. Schlesinger's copy of this transcript now resides in Widener Library, Harvard University, Cambridge, Mass.)

25. *ESS*, 1:xvii–xviii; Johnson to Stacy May, May 13, 1953, copy in Lubin papers.

26. Ruml to Seligman, May 23, 1927, S 3, Sub 5.5, B 51, F 540, LSRM papers; photostat of original contract in RG 1.1, S 200, B 330, F 3933, RF papers; overall sales record in "A Study for the Need for a New Cyclopedia of the Social Sciences," August 25, 1955, RG 1.1, S 200, B 330, F 3935, RF papers, all in RAC.

27. See Johnson to David Sills, June 22, 1964, Johnson papers, and Johnson, introduction to *International Encyclopedia of the Social Sciences*, 16 vols. (New York: Macmillan, 1968), 1:xi; E. R. A. Seligman, preface to *ESS*, 1:xxii.

28. Frankfurter to Seligman, March 28, 1926, reprinted in "A Memorandum on the Projected Encyclopedia of the Social Sciences," p. 15, B 38, F 319, Sage papers, RAC; Johnson to Keppel, October 31, 1930, copy in Seligman papers; unnamed author (probably Seligman), "Memorandum on the Projected Encyclopedia of the Social Sciences," p. 9, B 38, F 319, Sage papers, RAC.

29. Johnson to Agnes de Lima, December 29, 1929, Johnson papers.

30. Seligman, preface to *ESS*, 1:xxii; E. B. Wilson to Johnson, June 17, 1930, copy in RG 1.1, S 200, B 329, F 3924, RF papers, RAC. On the copyediting, see Johnson to Seligman, July 26, 1930, Seligman papers.

31. On the remuneration for contributions, see memorandum to *ESS* authors, Seligman papers; Johnson to Seligman, July 21, 1930, Seligman papers; Max Lerner, interview with the author, December 28, 1990; Johnson to David Sills, June 22, 1964, Johnson papers; Johnson to Seligman, July 21, 1930, and Seligman to Johnson, July 22, 1930, both in Seligman papers.

32. Day to Seligman, May 12, 1932, RG 1.1, S 200, B 330, F 3926, RF papers, RAC; Edwin Patterson, memorandum to Association of American Law Schools, copy in Seligman papers; photocopy of contract between Macmillan and Encyclopedia, Inc., RG 1.1, S 200, B 330, F 3933, RF papers, RAC.

33. Thomas B. Appelget diary entry, November 29, 1932, RG 1.1, S 200, B 330, F 3926, RF papers, RAC; Jones report, p. 6, B 3, F 32, Jones papers, RAC; Lerner interview; Appelget diary entry, February 3, 1933, RG 1.1, S 200, B 330, F 3927, RF papers, RAC; Seligman to Johnson, July 22, 1930, Seligman papers.

34. Jones report to RF, pp. 13, 6, 11, B 3, F 32, Jones papers, RAC; Alvin Johnson, *Pioneer's Progress* (New York: Viking, 1952), p. 314. Of the $1.25 million final cost, the Rockefeller Foundation contributed $625,000, the Carnegie Corporation $275,000, and the Russell Sage Foundation "a substantial amount," much of which was in loans later repaid. The publishers at Macmillan allowed an advance of $40,000. Edwin Patterson memo to American Association of Law Schools, n.d., Seligman papers.

35. On the smudge density index, see "A Study for the Need for a New Cyclopedic Treatment of the Social Sciences," August 25, 1955, RG 1.1, S 200, B 330, F 3935, RF papers, RAC. On rebinding, see Johnson, *Pioneer's Progress*, p. 313.

36. All quotations from *ESS* promotional pamphlet, B 38, F 319, Sage papers, RAC.

37. Guido de Ruggiero, "Positivism," in *ESS*, 12:265; Harold Laski, "Bureaucracy," in *ESS*, 2:70–74; Irwin Edman, "Art: Introduction," in *ESS*, 2:223; Horace Kallen, "James, William," in *ESS*, 8:369.

38. Editorial Staff, "Introduction I—The Development of Social Thought and Institutions," part 12: "War and Reorientation," in *ESS*, 1:204; Hermann Heller, "Political Science," in *ESS*, 12:207; Editorial Staff, "War and Reorientation," 1:194, 228.

39. Helen Everett, "Control, Social," in *ESS*, 4:346.

40. Ordway Tead, "Personnel Administration," in *ESS*, 12:88. Like Person, Tead had been an active Taylor Society member in the 1910s, and later was the editor for Gunnar Myrdal's *An American Dilemma* at Harper and Row.

41. Morris R. Cohen, "Method, Scientific," in *ESS*, 10:395; Benjamin Ginzberg,

"Science," in *ESS*, 13:602; Everett, "Control, Social," 4:345; H. S. Person, "Engineering," in *ESS*, 5:546.

42. For a fine discussion of the various meanings of *social control,* see Dorothy Ross, *The Origins of American Social Science* (New York: Cambridge University Press, 1991), pp. 247–56.

43. L. L. Bernard, "The Social Sciences as Disciplines," in *ESS*, 1:343; H. S. Person, "Industrial Education," in *ESS*, 7:697; Bernard, "Social Sciences," 1:348.

44. Elizabeth Todd, "Amateur," in *ESS*, 2:18–20; George Catlin, "Expert," in *ESS*, 6:10, 12.

45. Emil Lederer, "National Economic Planning," in *ESS*, 11:205; Walton Hamilton, "Organization, Economic," in *ESS*, 11:489; George Soule, "Stabilization, Economic," in *ESS*, 14:315.

46. Peter Rutkoff and William B. Scott, *New School: A History of the New School for Social Research* (New York: Free Press, 1986), pp. 67–83. According to Max Lerner, "Dewey? On the staff he was something of a hero, but not to me." Lerner's master's thesis had been written on Veblen, and Johnson himself was an active Veblenite as well. Lerner interview. Johnson to Lerner, October 8, 1948, Johnson papers.

47. Everett, "Control, Social," 4:347.

48. Cohen, "Method, Scientific," 10:395; Horace Kallen, "Pragmatism," in *ESS*, 12:307, 311.

49. John Dewey, "Logic," in *ESS*, 9:602; John Dewey, "Human Nature," in *ESS*, 7:536.

50. Dewey, "Logic," 9:602, 603.

51. Rutkoff and Scott, *New School,* p. 77; Day, "Trends of Social Science," p. 5.

52. Lindsay Rogers, "Politics," in *ESS*, 12:225; Hermann Heller, "Power, Political," in *ESS*, 12:305.

53. Max Lerner named Lewis Corey and Bernhard Stern in particular as Marxists among the assistant editors. Lerner interview.

54. *Recent Economic Changes in the United States: Report of the Committee on Recent Economic Changes of the President's Conference on Unemployment, Herbert Hoover, Chairman* (New York: McGraw-Hill, 1929).

55. Day and the Rockefeller administrators had been well informed about the project from the outset. Ogburn gave Day a summary of negotiations before the dinner meeting at which Hoover invited the social scientists to participate, feeling "very sorry to have a situation such that you can't very well accept an invitation to the Whitehouse [*sic*]." Max Mason, the president of the foundation, similarly had a clear idea of the project before the grant application came in. See Ogburn to Day, September 5, 1929, B 3, F 12, Ogburn papers, and Max Mason, diary entry of October 2, 1929, RG 1.1, S 200, B 326, F 3873, RF papers, RAC; the quotations are from the Mason diary.

56. My discussion of *Recent Social Trends* relies to a significant degree on Barry D. Karl, "Presidential Planning and Social Science Research: Mr. Hoover's Experts," *Perspectives in American History* 3 (1969): 347–412.

57. Ogburn, "A Note on Method," pp. 4–6, Merriam papers; Ogburn memo to staff of President's Research Committee on Recent Social Trends, January 25, 1932, copy in Merriam papers; transcript of committee meeting, February 13, 1932, p. 13, Odum papers; transcript of committee meeting, June 26, 1932, p. 2, Odum papers; transcript of committee meeting, June 23, 1932, p. 3, Odum papers. Pressure appears to have mounted on Ogburn, who admitted to Odum that "I find it increasingly hard to make decisions. I am so darned eager not to make any mistakes." Ogburn to Odum, May 13, 1930, Ogburn papers.

58. *Recent Social Trends in the United States*, Committee Findings (New York: McGraw-Hill, 1933), pp. xiii, xxxi; Wesley C. Mitchell, "Statistics and Government," reprinted in *The Backward Art of Spending Money and Other Essays*, by Wesley C. Mitchell (New York: McGraw-Hill, 1937), p. 49.

59. Transcript of committee meeting, June 29, 1932, pp. 3, 4, Odum papers.

60. Odum to Ogburn, March 9, 1932, Ogburn papers; transcript of committee meeting, December 17, 1932, p. 4, Odum papers.

61. John Dewey, review of *Recent Social Trends*, in *JD:LW*, 9:235; A. A. Berle, review of *Recent Social Trends*, *Saturday Review of Literature* 9 (1933): 535; Charles Beard, review of *Recent Social Trends*, *Yale Review* 22 (1933): 596. The *Saturday Review* frequently attacked scientism throughout the 1930s.

62. Pitirim Sorokin, "*Recent Social Trends*: A Criticism," *Journal of Political Economy* 41 (1933): 194–210; William F. Ogburn, "A Reply," ibid., pp. 210–21. See also Sorokin's "Rejoinder to Professor Ogburn's Reply," ibid., pp. 400–404.

63. Beard, review of *Recent Social Trends*, p. 596.

CHAPTER 8

1. Chicago fair committee to George Burgess, August 21, 1928, quoted in *A Century of Progress: Report of the President to the Board of Trustees*, March 14, 1936, p. 30; NRC response on p. 32.

2. My discussion of the Century of Progress relies on Robert W. Rydell, "The Fan Dance of Science: American World's Fairs in the Great Depression," *Isis* 76 (1985): esp. 525–35; Dawes is quoted on p. 527. Pupin, September 5, 1928, F 5–255, CoP papers.

3. Pupin quoted at Third Science Advisory Committee Meeting, March 28, 1930, p. 10, F 5–264; Maurice Holland, "Science Takes off Its High Hat," p. 2, F 5–248; Odum to John Sewell, January 13, 1931, F 11–213, all in CoP papers.

4. 1933 Century of Progress, *Official Guide Book of the Fair*, pp. 62, 10 (emphasis added).

5. Transcript of Third Science Advisory Committee Meeting, March 28, 1930, pp. 32–39, F 5–264, CoP papers.

6. *A Century of Progress*, p. 70; E. B. Wilson to Rudolph Clemen, December 5, 1930, Odum papers.

7. "Notes on the Development of the Social Science Division of A Century of Progress," October 10, 1931, p. 2, Odum papers.

8. Rydell, "Fan Dance of Science," p. 535.

9. "Introductory Statement Regarding A Century of Progress," December 1932, and "Notes on the Development of the Social Science Division of a Century of Progress," October 10, 1931, p. 2, both in Odum papers.

10. "Memorandum concerning the social science division, A Century of Progress, World's Fair, 1933," December 24, 1930, CoP papers.

11. SSRC board of directors minutes, September 13–16, 1932, p. 11, B 139, F 11, Merriam papers; "Trends in Social Research and the Application of Social Science to Society," B 3, F 8, Ogburn papers.

12. R. A. Clemen to "Major" [Lenox] Lohr, October 29, 1930, F 11–197, and "Memorandum: Social Science Exhibits," August 20, 1930, F 11–206, both in CoP papers.

13. Odum to John Sewell, December 8, 1931, and Odum memo, December 16, 1931, Odum papers.

14. Wilson to Crane, February 5, 1932, CoP papers; 1933 Century of Progress, *Official Guide Book of the Fair*, p. 60.

15. Odum to Helen Bennett, November 5, 1932, F 11–234, CoP papers; Daniel J. Singal, *The War Within: From Victorian to Modernist Thought in the South, 1919–1945* (Chapel Hill: University of North Carolina Press, 1982), p. 393 n. 67.

16. This account of van Kleeck draws in large measure from Guy Alchon, "Mary van Kleeck and Social-Economic Planning," *Journal of Policy History* 3 (1991): 1–23.

17. Van Kleeck to Morris Cooke, August 1931, Cooke papers.

18. Alchon, "Mary van Kleeck," pp. 9–11.

19. Mary van Kleeck, "World Social Economic Planning," in *World Social Economic Planning*, ed. Mary Fledderus (The Hague: [IRI], 1932), p. 569.

20. Van Kleeck to Raymond Fosdick, October 29, 1931, B 13, F 230, van Kleeck papers; IRI brochure included in van Kleeck to Morris Cooke, August 1931, Cooke papers.

21. Van Kleeck, "World Social Economic Planning," pp. 566–68; Mary van Kleeck, "Economic Planning for the United States," in *On Economic Planning*, ed. Mary Fledderus and Mary van Kleeck (New York: Covici Friede, 1935), p. 262.

22. Mary van Kleeck, "Soviet Planning," October 26, 1932, B 29, F 541, van Kleeck papers; William Spofford quoted in Alchon, "Mary van Kleeck," p. 14.

23. Mary van Kleeck, "Everybody's Interest in Social Economic Planning," radio broadcast, July 14, 1931, pp. 3, 2, B 28, F 521, van Kleeck papers; van Kleeck,

"Economic Planning for the United States," p. 242; Mary van Kleeck, "Our Illusions Regarding Government," May 1934, p. 10, B 24, F 491, van Kleeck papers.

24. Van Kleeck, "Our Illusions," p. 10; Mary van Kleeck, "Social Economic Planning in the United States," November 26, 1934, pp. 11–12, B 24, F 492, van Kleeck papers.

25. "Social Politics," interview with Mary van Kleeck, July 20, 1929, p. 1, B 24, F 488; Mary van Kleeck, "A Record of 35 Years of Fact-finding on the Social Aspects of Industry," 1940, B 99, F 1549; Mary van Kleeck, "Social Economic Planning," February 20, 1932, p. 2, B 24, F 489, all in van Kleeck papers.

26. Glenn Frank, "Back to the Spinning Wheel?," *Magazine of Business* 52 (1927): 412; Glenn Frank, "This Business Age," ibid., 53 (1928): 716. See also Glenn Frank, "Needed: A New Man of Business," ibid., 52 (1927): 565.

27. Bruce Sinclair, *A Centennial History of the American Society of Mechanical Engineers, 1880–1980* (Toronto: University of Toronto Press, 1980), p. 10.

28. Ralph Flanders, *Taming Our Machines* (New York: Richard Smith, 1931), pp. 17, 159, 168, 233.

29. Milton Nadworthy, *Scientific Management and the Unions* (Cambridge, Mass.: Harvard University Press, 1955), p. 106; Lyndall Urwick, *The Golden Book of Management* (London: Newman Neame, 1956), p. 227.

30. Warfield Webb, "Helping the Employees by Singing in the Plant," *Industrial Management*, June 1920, p. 508.

31. Membership figures taken from Cooke papers.

32. "Comment," *Bulletin of the Taylor Society* 14 (1928): 1; Person to Cooke, September 16, 1925, Cooke papers. Dennison's research associate summarized the situation in mid-1930: "A.S.M.E. and A.M.A. [American Management Association] have played closely together for last couple of years. T.S. [Taylor Society] has been left out on a limb, and H.P. [Harlow Person] not too happy about it." See American Management Association folder, Dennison papers.

33. Person to Cooke, April 15, 1927, and "Conference Dinner" seating chart, both in Cooke papers.

34. Keppel to Cooke, February 24, 1933, ibid.

35. On the Business Research Council, see the eponymous file in the Dennison papers. The council was supported by the American Association of Collegiate Schools of Business, the AEA, the American Federation of Labor, the American Management Association, the American Statistical Association, and the American Trade Association Executives.

36. Transcript, Taylor Society Fraternity Club dinner, April 28, 1927, pp. 2, 17, 9, Cooke papers.

37. Ibid., p. 14.

CHAPTER 9

1. Arland D. Weeks, "Will There Be an Age of Social Invention?," *Scientific Monthly* 35 (1932): 369.

2. RF board meeting, April 11–12, 1933, RG 3, S 910, B 1, F 1; "Social Sciences: Program and Policy," RG 3, S 910, B 2, F 12; E. E. Day, "Proposed Foundation Program in Economic Stabilization," September 14, 1931, RG 3, S 910, B 2, F 12, all in RF papers, RAC.

3. Robert Westbrook, "Tribune of the Technostructure: The Popular Economics of Stuart Chase," *American Quarterly* 32 (1980): 389–92, 395; John Chamberlain, "Stuart Chase: The Economic Wizard," *Modern Monthly* 7 (1933): 332; Stuart Chase, "Where Are All the Pre-war Radicals?," *Survey* 55 (1926): 564; Stuart Chase, *Government in Business* (New York: Macmillan, 1935), p. 281; Stuart Chase, *A New Deal* (New York: Macmillan, 1933), p. 232.

4. Stuart Chase, "The Technician and the Future," *World Tomorrow* 5 (1922): 367; Stuart Chase, *The Nemesis of American Business* (New York: Macmillan, 1931), pp. 103, 107, quoted in Westbrook, "Chase," p. 401; Stuart Chase, "A Ten Year Blueprint for America," *Harper's*, June 1931, p. 6; Chase, *A New Deal*, p. 177.

5. George Soule, *The Coming American Revolution* (New York: Macmillan, 1934), p. 304; George Soule, "Planning for Agriculture," *NR*, October 7, 1931, p. 206 (emphasis added).

6. George Soule, *A Planned Society* (New York: Macmillan, 1934), pp. 91, 141, 148, 152, 283.

7. Evelyn C. Brooks and Lee M. Brooks, "Five Years of Planning Literature," *Social Forces* 11 (1933): 430–65.

8. Henry Dennison, "The Need for the Development of Political Science Engineering," *APSR* 26 (1932): 242–43; Henry Dennison, *Organization Engineering* (New York: McGraw-Hill, 1931), p. 243. See also Kim McQuaid, "Henry S. Dennison and the 'Science' of Industrial Reform, 1900–1950," *American Journal of Economics and Sociology* 36 (1977): 79–98. On Filene's views, see E. A. Filene, *Successful Living in This Machine Age* (New York: Simon and Schuster, 1932). For Person's perspective, see Harlow Person, "The Approach of Scientific Management to National Planning," *Annals of the American Academy of Political and Social Science* 162 (1932): 18–26.

9. See Howard Odum, "The Planning of an American Region," *Plan Age*, February 1936. On Mumford, Odum, and regionalism, see John L. Thomas, "Lewis Mumford, Benton MacKaye, and the Regional Vision," in *Lewis Mumford: Public Intellectual*, ed. Thomas P. Hughes and Agatha C. Hughes (New York: Oxford University Press, 1990), pp. 66–99; and Allen Tullos, "The Politics of Regional Development: Lewis Mumford and Howard W. Odum," in Hughes and Hughes, *Mumford*, pp. 110–20.

10. "ESPA—Aims and Methods," *Plan Age*, December 1934, p. 12; Marion Hedges, "A Prologue to International Planning," *Plan Age*, September 1937, 15.

11. George Soule, "NESPA: December 1934–December 1940," *Plan Age*, November–December 1940, p. 290.

12. Leon Ardzrooni, who was at the New School in the early years, claimed that for his part, "Veblen maintained his faith and interest in Scott until the end." Ardzrooni, "Veblen and Technocracy," *Living Age*, March 1933, p. 40; William Akin, *Technocracy and the American Dream* (Berkeley: University of California Press, 1977), pp. 86, xiii.

13. Akin, *Technocracy*, pp. 86–87, 109; Harold Loeb, *Chart of Plenty* (New York: Viking, 1935), p. 165.

14. This fascinating document reveals much about the organization's philosophy. Behaviorist psychology, for example, is taken as scientific law: "Everything which Pavlov found to be true in the dog is true also in human beings" (p. 193). Anti-intellectualism also prevails: as an example of "complete functional incompetence," the authors offer "Professor Ortega y Gasset [who] is a Jesuit Professor of Philosophy at the University of Madrid, and, as such, so far as is publicly known, has never done anything of more importance in his entire life than to read books, talk, and write more books" (p. 205).

15. Ibid., pp. vii, 14, 224 (emphasis in original).

16. Beard to Raymond Fosdick, May 20, 1922, Fosdick papers, Princeton University, Princeton, N.J.; copy in DC 572, F 25, Beard papers (published with permission of Princeton University Libraries); Beard quoted in Bernard Borning, *The Political and Social Thought of Charles A. Beard* (Seattle: University of Washington Press, 1962), p. 70.

17. David Marcell, *Progress and Pragmatism: James, Dewey, Beard, and the American Idea of Progress* (Westport, Conn.: Greenwood, 1974), pp. 266, 278–79; David Marcell, "Charles Beard: Civilization and the Revolt against Empiricism," *American Quarterly* 21 (1969): 65–86.

18. Charles Beard, "Political Science," in *Research in the Social Sciences*, ed. Wilson Gee (New York: Macmillan, 1929), pp. 285, 286; Charles Beard, "Conditions Favorable to Creative Work in Political Science," *APSR* 24, supplement (1930): 29.

19. Charles Beard and William Beard, *The American Leviathan: The Republic in the Machine Age* (New York: Macmillan, 1930); Charles Beard, ed., *Whither Mankind* (New York: Longmans, Green, 1929); Charles Beard, ed., *Toward Civilization* (New York: Longmans, Green, 1930).

20. Charles Beard, introduction to Beard, *Whither Mankind*, p. 18; Charles Beard, epilogue to Beard, *Whither Mankind*, p. 404.

21. Charles Beard, introduction to Beard, *Toward Civilization*, p. 15; Beard to Flanders, October 27, 1929, DC572, F 22, Beard papers; Ralph Flanders, "The New Age and the New Man," in Beard, *Toward Civilization*, p. 25. Beard told Flanders

that "you have written a remarkable chapter which hits the target exactly in the center. It has a firm structure. It is full of ideas. It has verve in style. I like it immensely."

22. Charles Beard, "Summary—The Planning of Civilization," in Beard, *Toward Civilization*, pp. 297, 303; Howard Mumford Jones, "The Will for the Deed," *Saturday Review of Literature*, May 10, 1930, p. 1022.

23. Beard and Beard, *American Leviathan*, pp. 19, 615–16, 16–17.

24. Charles Beard, "Rushlights in Darkness," *Scribner's*, December 1931, p. 578; Charles Beard, "A Five-Year Plan for America," *Forum*, July 1931, p. 1.

25. Charles A. Beard and Mary Beard, *The Rise of American Civilization* (New York: Macmillan, 1927), 1:737; Charles Beard, "Written History as an Act of Faith," *American Historical Review* 39 (1934): 219–31.

26. Charles Beard, introduction to *Towards Technocracy*, by Graham Laing (Los Angeles: Angelus Press, 1933), p. x; Charles Beard, "Limitations in the Application of Social Science Implied in *Recent Social Trends*," *Social Forces* 7 (1933): 510.

27. Charles Beard, review of Harold Loeb, *The Chart of Plenty*, NR, March 20, 1935, p. 164. See also "Twenty-Five on Technocracy," *Common Sense*, February 2, 1933, pp. 8–10.

28. Charles A. Beard and Mary Beard, *The Rise of American Civilization*, rev. ed. (New York: Macmillan, 1933), p. 837.

29. An outpouring of compelling Dewey scholarship has appeared in the last decade, and I hasten to direct the reader to these much fuller accounts to which I owe much: James T. Kloppenberg, *Uncertain Victory: Social Democracy and Progressivism in European and American Thought, 1870–1920* (New York: Oxford University Press, 1986); Richard Rorty, *Consequences of Pragmatism* (Minneapolis: University of Minnesota Press, 1982); Dorothy Ross, *The Origins of American Social Science* (New York: Cambridge University Press, 1991); Cornel West, *The American Evasion of Philosophy: A Genealogy of Pragmatism* (Madison: University of Wisconsin Press, 1989); and Robert Westbrook, *John Dewey and American Democracy* (Ithaca: Cornell University Press, 1991).

30. Westbrook, *Dewey*, pp. 448–52.

31. Advertisements in *Common Sense*, March 30, and February 16, 1933; "*Common Sense* Affirms," *Common Sense*, December 1937, p. 1. For more on Bingham, see Donald Miller, *The New American Radicalism* (Port Washington, N.Y.: Kennikat, 1979).

32. John Dewey, *Liberalism and Social Action*, in *JD:LW*, 11:51–52.

33. John Dewey, *Ethics*, in *JD:MW*, 5:268, quoted in Westbrook, *Dewey*, p. 158; Westbrook, *Dewey*, p. 166.

34. John Dewey, "I Believe," in *JD:LW*, 14:91. See also Timothy Kaufman-Osborn, "Pragmatism, Policy Science, and the State," *American Journal of Political Science* 29 (1985): 828.

35. Westbrook, *Dewey*, p. 526.

36. John Dewey, *Logic: The Theory of Inquiry*, in *JD:LW*, 12:84.

37. John Dewey, *Art as Experience*, in *JD:LW*, 10:347.

38. John Dewey, *Lectures on Psychological and Political Ethics: 1898* (New York: Hafner Press, 1976), pp. 441–44, quoted in Westbrook, *Dewey*, p. 93; Westbrook, *Dewey*, p. 108.

39. John Dewey, *Reconstruction in Philosophy*, in *JD:MW*, 12:121; John Dewey, review of Walter Lippmann, *An Inquiry into the Principles of the Good Society*, in *JD:LW*, 11:495. On cultural lag, see Dewey, *Liberalism and Social Action*, 11:53.

40. John Dewey, "The Economic Basis of the New Society," in *JD:LW*, 13:320.

41. John Dewey, "Pragmatic Acquiescence," in *JD:LW*, 3:145.

42. John Dewey, *Individualism Old and New*, in *JD:LW*, 5:107. On optimism, see Louis Menand, "The Real John Dewey," *New York Review of Books*, June 25, 1992, p. 55: "For unlike almost every other writer of his time [and like the other technocratic reformers, I would add], Dewey did not regard modern life as a deprivation, a cultural wound which the absence of traditional kinds of religious and civic faith made it impossible to heal."

43. John Dewey, "Social Science and Social Control," in *JD:LW*, 6:66.

44. Dewey, *Individualism Old and New*, 5:86.

45. John Dewey, *Experience and Nature*, in *JD:LW*, 1:128.

46. Dewey, *Individualism Old and New*, 5:87.

47. Dewey, *Liberalism and Social Action*, 11:38.

48. Rorty, *Consequences of Pragmatism*, p. 56 n. 38.

CHAPTER 10

1. Richard Adelstein, "'The Nation as an Economic Unit': Keynes, Roosevelt, and the Managerial Ideal," *Journal of American History* 78 (1991): 160–87.

2. Otis Graham, *Toward a Planned Society: From Roosevelt to Nixon* (New York: Oxford University Press, 1976), p. 25.

3. Mordecai Ezekiel, "Economic Philosophy of the New Deal," *Journal of Business*, May 1936, p. 4; Lewis Lorwin, "Planning in a Democracy," *Plan Age*, February 1934, p. 6; Lewis Lorwin, "Some Political Aspects of Economic Planning," *APSR* 26 (1932): 27–28.

4. On the NRA, see Ellis W. Hawley, *The New Deal and the Problem of Monopoly* (Princeton: Princeton University Press, 1964).

5. Charles F. Roos, *NRA Economic Planning* (Bloomington, Ind.: Principia, 1937), p. 472.

6. George Galloway, *Planning for America* (New York: Holt, 1941), pp. 645, 15, 64.

7. Barry D. Karl, *Charles E. Merriam and the Study of Politics* (Chicago: University of Chicago Press, 1974), p. 248. The standard sources on the NRPB are Marion Clawson, *New Deal Planning* (Baltimore: Johns Hopkins University Press, 1981),

and Philip Warken, *A History of the National Resources Planning Board, 1933–1943* (New York: Garland, 1979). Over the following decade, the National Planning Board (1933–34) gave way to the National Resources Board (1934–35) before the National Resources Committee (1935–39) was formed in the midst of Roosevelt's reorganization of the White House offices. The NRPB, which survived until 1943, was the final manifestation, and it is by this abbreviation that all of the different incarnations will be noted here.

8. Delano to Walter Lippmann, August 2, 1935, Delano papers; Frederic Delano, "The Economic Implications of National Planning," *Proceedings of the American Philosophic Society* 74 (1934): 21–28; Delano to Stacy May, March 13, 1935, copy in B 39, F 329, Sage papers, RAC.

9. For a bibliography of the NRPB's studies, see Clawson, *New Deal Planning*, pp. 326–47.

10. National Planning Board, Final Report, 1933–34, "A Plan for Planning," pp. 18, 34.

11. Ibid., pp. 32, 33.

12. Ibid., p. 31.

13. "B. Ruml's Memorandum," December 1933, RG 3, S 910, B 2, F 13, RF papers, RAC. On national income and government spending, see "Miami memorandum," April 6, 1938, S II, B 1, F 26, Ruml papers, and untitled memorandum to National Resources Committee, April 30, 1938, B 186, F 8, Merriam papers.

14. Wesley C. Mitchell, "The Application of Economic Knowledge," in *The Obligation of Universities to the Social Order* (New York: New York University Press, 1933), p. 174; NBER 1938 annual report, p. 31; Wesley C. Mitchell, "The Social Sciences and National Planning," in *The Backward Art of Spending Money and Other Essays*, by Wesley C. Mitchell (New York: McGraw-Hill, 1937), p. 102.

15. Wesley C. Mitchell, "Economics in a Changing Social Order," NBC broadcast, October 4, 1934, p. 4, in Tugwell papers; Wesley C. Mitchell, "Supplementary Statement," in *Economic Reconstruction: Report of the Columbia University Commission* (New York: Columbia University Press, 1934), p. 82.

16. Wesley C. Mitchell, Messenger Lecture #1, Cornell University, February 18, 1935, p. 15, Wesley C. Mitchell papers; NBER, 1938 annual report, p. 31; Alvin Hansen quoted in Arthur Burns, ed., *Wesley Clair Mitchell: The Economic Scientist* (New York: NBER, 1952), p. 319.

17. Charles E. Merriam, *A Prologue to Politics* (Chicago: University of Chicago Press, 1939), p. 92; "Planning and Science," undated address, pp. 30, 44, 41, B 295, F 10, Merriam papers (emphasis added).

18. Charles E. Merriam, "The National Resources Planning Board: A Chapter in America's Planning Experience," *APSR* 38 (1944): 1076; Charles E. Merriam, "The National Resources Planning Board," *Public Administration Review* 1 (1940): 121.

19. Roy Talbert, *FDR's Utopian: Arthur Morgan of the TVA* (Jackson: University Press of Mississippi, 1987), pp. 53, 196; Morgan called Bellamy "almost a modern

Leonardo da Vinci" and grouped him with Freud, Emerson, Hawthorne, and Thoreau.

20. Ibid., esp. pp. 52, 58; Thomas McCraw, *Morgan vs. Lilienthal: The Feud within the TVA* (Chicago: Loyola University Press, 1970), esp. p. 64.

21. Arthur Morgan, *The Long Road* (Washington, D.C.: National Home Library Foundation, 1936), pp. 30, 29, 41.

22. Morgan, *The Long Road*, pp. 1, 38, 74; Roy Talbert, Jr., ed., "Arthur E. Morgan's Ethical Code for the Tennessee Valley Authority," *East Tennessee Historical Society Publications* 40 (1968): 123, 122 (emphasis added).

23. Talbert, "Morgan's Ethical Code," p. 122.

24. Ibid., pp. 123–27.

25. Talbert, *FDR's Utopian*, p. 129.

26. Ellis Hawley, "Economic Inquiry and the State in New Era America," in *The State and Economic Knowledge: The American and British Experiences*, ed. Mary Furner and Barry Supple (New York: Cambridge University Press, 1990), p. 322. The USDA graduate school appears to have been a center for many discussions related to social engineering; whether this had anything to do with Tugwell has not been determined. Lecturers such as the social planner David Cushman Coyle, Morris R. Cohen, the biologists John C. Merriam and Walter Cannon, the engineer Dexter Kimball, and Harry Elmer Barnes came as part of lecture series, usually ten weeks long, called "Science: Its History, Philosophy, and Relation to Democracy," which addressed social problems, managerial skills, and other topics closely associated with technology and its social context.

27. Rexford Tugwell, *To the Lesser Heights of Morningside* (Philadelphia: University of Pennsylvania Press, 1982), p. 180; Rexford Tugwell, *The Battle for Democracy* (New York: Columbia University Press, 1935), p. 260. See also Steven Kesselman, *The Modernization of American Reform* (New York: Garland, 1979), esp. pp. 370–415, and Bernard Sternsher, *Rexford Tugwell and the New Deal* (New Brunswick, N.J.: Rutgers University Press, 1964).

28. Rexford Tugwell, "Social Objectives in Education," September 1, 1932, pp. 96–97, copy in speech and writings file (emphasis added), and Tugwell to George Soule, October 22, 1932, general correspondence file, both in Tugwell papers.

29. Rexford Tugwell, "Social Objectives in Education," in *Redirecting Education*, ed. Leon Keyserling and Rexford Tugwell (New York: Columbia University Press, 1934), p. 100; Day to Tugwell, January 21, 1932, B 368, F 3831, GEB papers, RAC.

30. Tugwell, *Battle*, pp. 14, 245.

31. Ibid., p. 243; Rexford Tugwell, *The Industrial Discipline and the Governmental Arts* (1933; reprint, New York: Arno, 1977); John Dewey, review of Tugwell, *Industrial Discipline*, in *JD:LW*, 8:364–66.

32. Rexford Tugwell, "The Superpolitical," *Journal of Social Philosophy* 5 (1940): 97–114.

CHAPTER 11

1. Lewis Mumford, *Findings and Keepings* (New York: Harcourt, Brace, Jovanovich, 1975), p. 376. See also Donald Miller, *Lewis Mumford: A Life* (New York: Weidenfeld and Nicholson, 1989).

2. Lewis Mumford, *Sketches from Life* (Boston: Beacon Press, 1982), pp. 251, 221, 187.

3. Lewis Mumford to Van Wyck Brooks, January 20 and March 26, 1933, in *The Van Wyck Brooks–Lewis Mumford Letters: The Record of a Literary Friendship*, ed. Robert Spiller (New York: Dutton, 1976), pp. 90, 92; Mumford to Catherine Bauer, June 4, 1933, in Mumford, *My Works and Days: A Personal Chronicle* (New York: Harcourt, Brace, Jovanovich, 1979), p. 309; Mumford to Brooks, June 21, 1933, in Spiller, *Brooks–Mumford Letters*, p. 95.

4. Herbert Read, review of Lewis Mumford, *Technics and Civilization*, *Yale Review* 24 (1934): 173.

5. By "the machine," Mumford meant the entire cultural complex surrounding modern industry. Lewis Mumford, *Technics and Civilization* (New York: Harcourt, Brace, 1934), pp. 26–27, 89, 94.

6. Ibid., pp. 396, 367, 421, 46, 408.

7. Ibid., pp. 370, 322, 281. On Mumford and Dewey, see Robert Westbrook, "Lewis Mumford, John Dewey, and the 'Pragmatic Acquiescence,'" in *Lewis Mumford: Public Intellectual*, ed. Thomas P. Hughes and Agatha C. Hughes (New York: Oxford University Press, 1990), pp. 301–22.

8. Mumford, *Technics and Civilization*, pp. 6, 363, 435.

9. Ibid., p. 417.

10. Ibid., p. 389. See also pp. 380, 410.

11. Lewis Mumford, *The Story of Utopias* (New York: Boni and Liveright, 1922), p. 276 (emphasis added); Mumford, *Technics and Civilization*, p. 420 (emphasis added); Casey Blake, "Lewis Mumford: Values over Technique," *democracy*, Spring 1983, p. 129.

12. Mumford, *Utopias*, p. 7.

13. Walter Lippmann, *A Preface to Morals* (New York: Macmillan, 1929), p. 283; Walter Lippmann, *A New Social Order* (New York: John Day, 1933), pp. 21, 15.

14. Walter Lippmann, *The Method of Freedom* (New York: Macmillan, 1934), pp. 66, 74; Lippmann to Gerald Johnson, May 18, 1928, in *Public Philosopher: Selected Letters of Walter Lippmann*, ed. John Morton Blum (New York: Ticknor and Fields, 1985), p. 220. On planning, see Lippmann to Charles Beard, September 8, 1925, Lippmann papers.

15. Lippmann to Arthur N. Holcombe, September 22, 1936, in Blum, *Public Philosopher*, p. 354; Lippmann to Lionel C. Robbins, March 24, 1937, in ibid., p. 357; Lippmann to Hayek, March 12, and Hayek to Lippmann, April 6, 1937, both in Lippmann papers.

16. Walter Lippmann, *The Good Society* (Boston: Little, Brown, 1937), esp. p. 16.

17. Ibid., pp. 33, 52, 100–101.

18. Ibid., pp. 384–87.

19. Newell Sims, review of Walter Lippmann, *The Good Society*, *American Sociological Review* 3 (1938): 110; Ralph Barton Perry, "The Liberal State," *Yale Review* 27 (1937): 404; Max Lerner, "Lippmann Agonistes," *Nation*, November 27, 1937, p. 590; Max Lerner, *It Is Later Than You Think* (New York: Viking, 1939), p. 139; Charles Merriam, review of Walter Lippmann, *The Good Society*, APSR 53 (1938): 129–34; Lewis Mumford, "Mr. Lippmann's Heresy Hunt," *NR*, September 29, 1937, pp. 219–20.

20. Beard to Maverick, September 8, 1937, copy in Beard papers. See also Brian VanDeMark, ed., "Beard on Lippmann: The Scholar vs. the Critic," *New England Quarterly* 59 (1986): 402–5.

21. John Dewey, "Liberalism in a Vacuum," *Common Sense*, December 1937, reprinted in *JD:LW*, 11:489–95.

22. Alexander Goldenweiser, "The Nature and Tasks of the Social Sciences," *Journal of Social Philosophy* 2 (1936): 5, 9, 26.

23. Howard Odum, "Facing Our Social Problems," lecture to graduate school, U.S. Department of Agriculture, March 31, 1939, pp. 20, 22, and Odum to T. J. Wilson, April 24, 1939 (approx.), both in Odum papers.

24. Howard Odum, *American Social Problems* (New York: Holt, 1939), pp. vi, 27, 299, 19; Odum to Harvard Alumni Association, June 22, 1939, p. 9, Odum papers.

25. Edward Purcell, *The Crisis of Democratic Theory: Scientific Naturalism and the Problem of Value* (Lexington: University Press of Kentucky, 1973), p. 196; Louis Wirth, internal report on SSRC history, activities, and policies, August 1937, pp. 1, 2, RG 1.1, S 200, B 401, F 4747, RF papers, RAC.

26. Wirth report, pp. 108, 104–5.

27. Hutchins quoted in Louis Wirth, ed., *Eleven Twenty-Six: A Decade of Social Science Research* (Chicago: University of Chicago Press, 1940), p. 3.

28. John Vansickle to Raymond Fosdick, January 4, 1937, and Simon Kuznets to Joseph Willets, August 1, 1939, both in RG 3, S 910, B 1, F 4, RF papers, RAC.

29. Day to E. B. Wilson, December 3, 1937, B 27, F 5, Day papers; "Conference on Social Education: Explanatory Statement," January 10, 1939, copy in ibid.

30. "Explanatory Statement."

31. Robert Lynd, *Knowledge for What?: The Place of Social Science in American Culture* (1939; reprint, Princeton: Princeton University Press, 1948), pp. 113, 125, 129 (emphasis in original).

32. Ibid., pp. 144–45, 166–67.

33. Ibid., pp. 185, 200.

34. Ibid., p. 213.

35. Charles Beard, review of Lynd, *Knowledge for What?*, APSR 33 (1939): 712;

Stuart Chase, review of Lynd, *Knowledge for What?*, *New York Herald-Tribune*, July 2, 1939; Max Lerner, "The Revolt against Quietism," *NR*, July 5, 1939, p. 258.

36. Crane Brinton, "What's the Matter with Sociology?," *Saturday Review of Literature*, May 6, 1939, pp. 3, 14.

37. Lynd to Robert MacIver's class at Columbia, 1938, Lynd papers; Lynd to Mitchell, May 30, 1944, Wesley C. Mitchell papers.

38. Wesley C. Mitchell, "Social Planning," discussion with Bureau of Educational Experiments, March 17, 1934, and Wesley C. Mitchell, Messenger Lecture #4, p. 13, March 1, 1935, both in Wesley C. Mitchell papers.

39. Wesley C. Mitchell, Columbia Economics Club discussion of Lynd, *Knowledge for What?*, May 2, 1939, transcript in ibid.

40. Ibid., p. 20. In the NBER 1940 annual report, pp. 8–15, Mitchell wrote that "it is no wonder that our procedure is criticized by many impatient people and by some thoughtful ones."

41. Ruml to Merriam, July 13, 1951, S 1, B 3, F 11, Ruml papers.

42. "The Road to Serfdom," radio broadcast of University of Chicago Round Table, April 22, 1945, transcript in B 284, F 18, Merriam papers.

43. Ibid., pp. 9, 11.

44. Merriam to Ruml, May 5, 1945, S 1, B 3, F 11, Ruml papers.

CONCLUSION

1. At the ten-year anniversary celebration of the Chicago social science research building, Beardsley Ruml, who largely made that building possible, returned. He spoke at the proceedings but was less committed to a social engineering perspective than he had been at the opening festivities, insisting that "a 'social inventor'. . . is certainly not a scientist." See Louis Wirth, ed., *Eleven Twenty-Six: A Decade of Social Science Research* (1940; reprint, New York: Arno, 1970), p. 290.

2. *Illustrated London News*, April 29, 1939, p. 707; *New York World's Fair Souvenir Book*; Brock Brower, "1939: Now *That* Was a World's Fair!," *Esquire*, April 1964, p. 65.

3. Gerard Wendt quoted in Joseph P. Cusker, "The World of Tomorrow: Science, Culture, and Community at the New York World's Fair," in *Dawn of a New Day: The New York World's Fair, 1939–40* (New York: Queens Museum–New York University Press, 1980), p. 12; H. G. Wells, "World of Tomorrow," *New York Times*, World's Fair ed., March 5, 1939, p. 5; *New York World's Fair Program*, p. 86.

4. Grover Whalen, "What the Fair Means to Business and Industry," *New York World's Fair Bulletin*, June 1937, p. 3; Whalen quoted in *New York World's Fair Souvenir Book*; "*Life* Goes to the New York World's Fair," *Life*, July 3, 1939, p. 57; *New York World's Fair Official Guide Book*, 3d ed., pp. 45, 53.

5. Helen Harrison, introduction to *Dawn of a New Day*, pp. 1–3.

6. For one explanation of the cultural implications of American technology in this period, see Thomas P. Hughes, *American Genesis: A Century of Invention and Technological Enthusiasm, 1870–1970* (New York: Viking, 1989).

7. On technology and liberalism, see Yaron Ezrahi, *The Descent of Icarus: Science and the Transformation of Contemporary Democracy* (Cambridge, Mass.: Harvard University Press, 1990).

INDEX

Gantt, Henry L., 40, 52, 54, 59–62, 63, 65, 95, 106, 107, 210, 213

Gay, Edwin, 42, 52, 95–96, 99, 100, 108, 119, 135, 156, 203, 204, 243

Giddings, Franklin H., 88, 131–32, 193

Gilbreth, Frank, 40, 42, 43, 52, 66, 95

Gilbreth, Lillian, 40, 217

Goldenweiser, Alexander, 175, 267

Gulick, Luther H., 44, 51, 52

Hall, A. B., 157, 164

Hamilton, Walton, 132, 156, 175

Harrison, Shelby, 156, 180, 188

Hart, Hornell, 152, 184

Harvard University, 23, 42, 47, 69, 72, 79

Hawley, Ellis W., 116, 117, 248

Hayek, Friedrich von, 263, 268, 278–79, 281

Hoover, Herbert, 140, 147, 195, 199, 203, 218, 225, 281, 282, 286; and engineering, 21, 33–36, 41, 110, 120–24, 200; and wartime food issues, 79, 95–102 passim, 110, 114; and presidential campaign of 1920, 104, 111–12; as secretary of commerce, 108, 114, 116–18, 134, 135, 179, 282; political philosophy of, 112–28; presidency of, 118–20, 134, 179–83, 232; postpresidential career, 127–28

House, Edward, 85, 96, 98

Hoxie, Robert, 16, 64

Hunt, Edward Eyre, 79, 180, 183, 203

Hutchins, Robert M., 269–70

Inquiry, the, 96–97, 103, 131

Institute for Governmental Research (IGR), 85, 87–88, 97, 133, 135. *See also* Brookings Institution

International Industrial Relations Institute (IRI), 193–99, 234

James, William, 14, 69, 83, 91, 161, 171

Johnson, Alvin S., 72, 83, 104, 156, 165–79, 184, 210

Johnson, Hugh, 234

Jones, Howard Mumford, 218

Jones, Mark, 170–71

Kallen, Horace, 172, 175, 176

Karl, Barry, 115, 235

Kellogg, Paul U., 86, 135, 184, 203

Kendall, Henry, 64–65, 95

Keppel, Frederick, 133, 135, 156, 168, 178, 203, 204

Keynes, John M., 110

Kloppenberg, James, 78

Knight, Frank, 161–65, 258

Knoeppel, C. E., 63, 82, 200

Kuznets, Simon, 270

Labor Bureau, U.S., 209

Lane, Franklin K., 105–6

Laski, Harold, 71, 157–58, 164, 165, 171, 310 (n. 5)

League for Independent Political Action, 222, 225

Lenin, V. I., 19, 55

Lerner, Max, 19, 170, 172, 175, 265, 266, 275

Liberalism and liberals, 8, 172–73, 177–78, 261, 271

Lilienthal, David, 243

Lippmann, Walter, 7, 19, 22, 47, 68–75, 78, 79, 81, 83, 95, 96–98, 104, 108–9, 112, 118, 124, 126, 203, 209, 225, 226, 243, 256, 261–66, 268

Loeb, Harold, 213, 221, 222

Lorwin, Lewis, 233–34

Lowell, A. Lawrence, 25–26

Public administration, 6, 7, 76, 84–86, 87, 140

Rautenstrauch, Walter, 213
Recent Economic Changes, 119, 179
Recent Social Trends, 116, 119–20, 144, 147, 150, 151, 155, 179–84, 192, 237
Resettlement Administration, 251
Riesman, David, 107
Rockefeller Foundation (and associated philanthropies), 7, 68, 87–88, 99, 102, 119, 133–34, 135–44, 178, 179, 184, 186, 192, 196, 203, 205, 208, 236, 239, 249–50, 252, 266, 270–72, 281, 313 (n. 55); Laura Spelman Rockefeller Memorial (LSRM), 137–42, 153, 155–84, 201, 277, 308 (n. 17); Social Science Division, 139, 141–44, 155–84, 208, 266; Spelman Fund, 144; General Education Board, 249–50, 270–72
Roosevelt, Franklin D., 7, 22, 94, 112, 147, 182, 231, 232, 233, 234, 237, 243, 245, 251, 261
Roosevelt, Theodore, 22–23, 37, 44, 51, 63, 69–70, 73, 77
Rorty, Richard, 231
Ross, Dorothy, 130, 289 (n. 5)
Ruml, Beardsley, 99, 108, 137–43, 155–65, 167, 178, 202, 208, 236–43, 260, 266, 277–78, 279, 325 (n. 1)
Russell Sage Foundation, 68, 134–35, 156, 195–98, 203
Rydell, Robert, 189

Science, conceptions of, 158–65, 173–79, 186–92, 216, 227–31
Scientific management, 7, 36–55, 146, 172, 195–98. *See also* Taylor, Frederick Winslow; Taylor Society

Scott, Howard, 213, 222, 318 (n. 12)
Seligman, E. R. A., 52, 165–79, 184
Shotwell, James T., 96, 131
Singal, Daniel, 192
Small, Albion, 27–28, 131, 271
Social control, 5, 9, 24, 93, 155, 156, 172–79, 228, 271, 274, 289 (n. 5)
Social Science Research Council (SSRC), 7, 132, 140, 144, 147, 155–65, 169, 180, 182, 184, 188–92, 201, 204, 208, 237, 252, 268–69, 270, 272, 273
Social sciences, 23–29, 84–89, 118–20, 129–32, 144–54, 155–84, 186–92, 266–79. *See also* specific disciplines
Social work and social workers, 84, 86–87, 134, 140
Society to Promote the Science of Management. *See* Taylor Society
Sociology, 27–28, 84, 88–89, 130, 131–32, 172, 275
Sorokin, Pitirim, 184
Soule, George, 83, 134, 172, 175, 209, 210–11, 212, 222, 238, 247, 263
Soviet Union, 55, 172, 197, 210, 264
Spengler, Oswald, 19, 217, 221
Steinmetz, Charles P., 9, 22, 36, 55–59, 61, 81, 282
Sullivan, Mark, 121, 180
Sumner, William Graham, 6, 14, 131
Swope, Gerard, 211, 232

Tarbell, Ida, 95, 112, 195
Taylor, Frederick Winslow, 7, 31, 34, 35, 36, 40–44, 51, 53–55, 57, 59, 65, 76, 115, 133, 195, 196, 197, 200, 213, 248, 295 (n. 16)
Taylor Society, 42, 51–52, 63–67, 79, 95, 134, 172, 195, 200–206, 211, 247, 316 (n. 32)
Tead, Ordway, 64, 76, 172, 173